# BASIC
# ELECTRONIC
# AND
# ELECTRICAL
# DRAFTING

JAMES D. BETHUNE

*Boston University*
*College of Engineering*
*Boston, Massachusetts*

# BASIC ELECTRONIC AND ELECTRICAL DRAFTING

## Second Edition

PRENTICE-HALL, INC., Englewood Cliffs, New Jersey 07632

*Library of Congress Cataloging in Publication Data*

Bethune, James D.,
 Basic electronic and electrical drafting.

  Includes index.
  1. Electronic drafting. 2. Electric drafting.
 I. Title.
 TK7866.B49 1985      621.3'022'1      84-13358
 ISBN 0-13-060336-8

Editorial/production supervision
 and interior design: Ellen Denning
Cover design: Lundgren Graphics, Ltd.
Manufacturing buyer: Gordon Osbourne
Page layout: Meg Van Arsdale

Printed in the United States of America

10  9  8  7  6  5  4  3  2  1

ISBN 0-13-060336-8 01

PRENTICE-HALL INTERNATIONAL, INC., *London*
PRENTICE-HALL OF AUSTRALIA PTY. LIMITED, *Sydney*
EDITORA PRENTICE-HALL DO BRASIL, LTDA., *Rio de Janeiro*
PRENTICE-HALL CANADA INC., *Toronto*
PRENTICE-HALL HISPANOAMERICANA, S.A., *Mexico*
PRENTICE-HALL OF INDIA PRIVATE LIMITED, *New Delhi*
PRENTICE-HALL OF JAPAN, INC., *Tokyo*
PRENTICE-HALL OF SOUTHEAST ASIA PTE. LTD., *Singapore*
WHITEHALL BOOKS LIMITED, *Wellington, New Zealand*

To KENDRA

# CONTENTS

## 13  COMPUTER APPLICATIONS TO ELECTRONIC AND ELECTRICAL DRAFTING    264

## APPENDICES    271

## INDEX    283

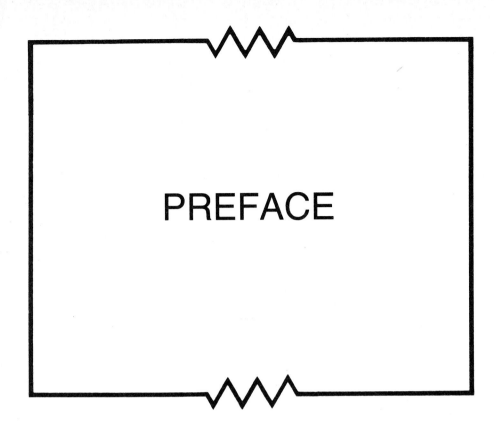

# PREFACE

I would like to thank everyone who helped make the initial edition of this book a success and further thank those who took the time to send in comments and corrections. I have tried to incorporate as many of these suggestions as possible and hope you will find the second edition more complete and accurate than the first.

There are three major areas of change: printed circuit drawings, ICs, and computer applications. Chapter 6, on printed circuit drawings, has been completely rewritten. The chapter includes space allocation drawings, master layouts, tape masters, single- and double-sided board drawing requirements, soldering masks, and component outline drawings. The chapter now includes over fifty illustrations as well as seven new exercise problems.

Chapter 7 is new and deals with integrated circuits: in particular, the design and layout of masks, together with a brief explanation of how ICs are manufactured.

ICs have been added throughout the book. ICs are integrated into the chapters on symbols, schematic diagrams, and logic diagrams. Chapter 6 uses ICs in three of the four examples that show the development of printed circuit boards.

Chapter 13 is a survey discussion of computer applications to electronic and electrical drafting. A brief overview of how various systems operate is given, and large and small systems are compared.

Chapter 1, which covers basic technical drawing fundamentals, has been expanded to include working drawings. The quantity of exercise problems has been doubled to assist instructors who must teach both basic drawing and electronic drawing within the same course.

Several people deserve special recognition for their help in preparing the book: my editors, Dave Boelio and Greg Burnell, who were both encouraging and helpful in the book's development; Ellen Denning, the production editor, who did a good job of making sense out of the pile of drawings and typed pages that I submitted; Sally Fischer, who did the typing; Bonnie Kee and Keith Clinkscale, who helped with the artwork; and Professor Maveretic of Boston University, who gave me excellent technical advice on many of the new sections.

If you have any comments or suggestions, please write to me in care of Prentice-Hall. Your comments on the first edition were helpful and are appreciated.

JAMES D. BETHUNE
*Boston, MA*

# BASIC
# ELECTRONIC
# AND
# ELECTRICAL
# DRAFTING

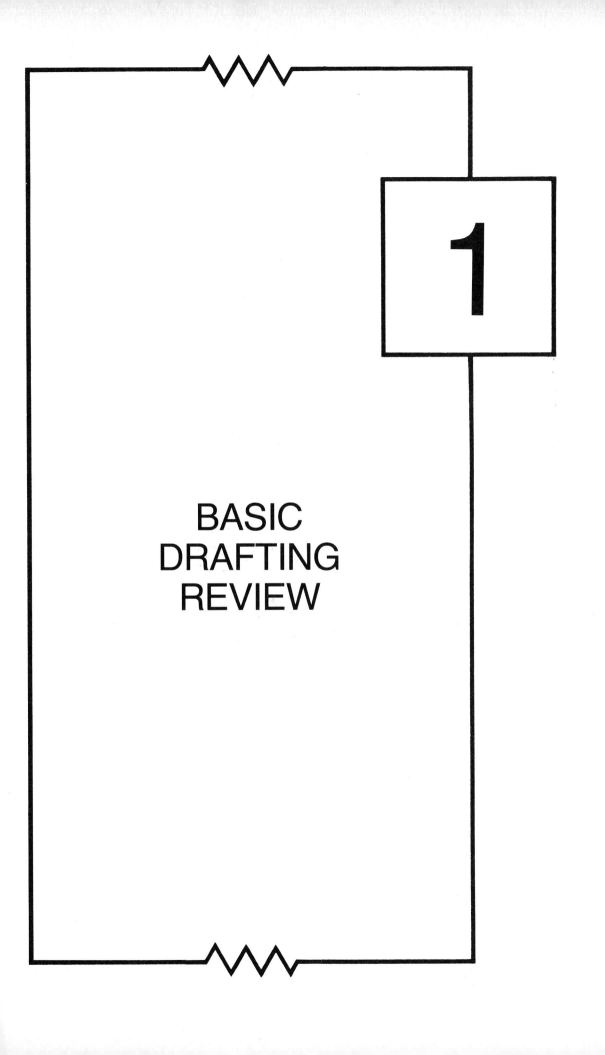

# 1

# BASIC
# DRAFTING
# REVIEW

## 1-1 INTRODUCTION

This chapter is a review of the fundamentals of basic drafting. It represents only a brief outline of the subject matter. The student who requires a more extensive review is referred to one of the following drafting texts:

Bethune, J. D., *Essentials of Drafting*. Englewood Cliffs, N.J.: Prentice-Hall, Inc., 1977.

Luzadder, W. J., *Fundamentals of Engineering Drawing*, 7th ed. Englewood Cliffs, N.J.: Prentice-Hall, Inc., 1977.

## 1-2 DRAFTING EQUIPMENT

Figure 1-1 pictures most of the drafting equipment needed to prepare electronic drawings and diagrams: a T-square, a 30-60-90 triangle, a 45-45-90 triangle, a lead holder, an eraser, an erasing shield, a protractor, a scale, drawing tape, a circle template, a compass, and an electronic symbol template. In the upper right-hand corner of the drawing board are a lead sharpener and a Styrofoam block. The Styrofoam block is used to remove excessive graphite from a freshly sharpened lead.

Figure 1-2 shows a drafting machine, which replaces the T-square, triangles, scale, and protractor. Operating instructions for drafting machines are supplied by the manufacturers.

Figure 1-3 shows how a T-square and triangles are used as guides when drawing horizontal and vertical lines. Remember to pull the lead

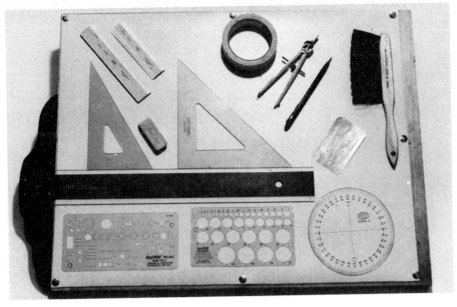

**FIGURE 1-1** Basic drawing equipment needed to prepare electronic drawings.

**FIGURE 1-2** Using a drafting machine. (*Courtesy of Teledyne Post, Des Plaines, IL 60016.*)

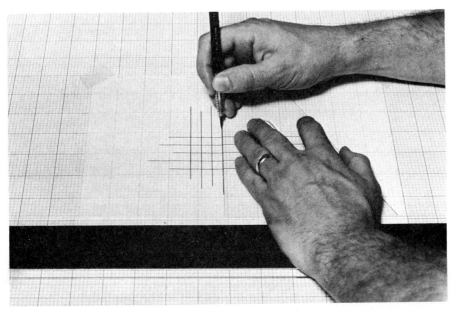

**FIGURE 1-3** Using a T-square and a triangle. The T-square is used as a guide for drawing horizontal lines and the T-square and triangle are used as shown as a guide for drawing vertical lines.

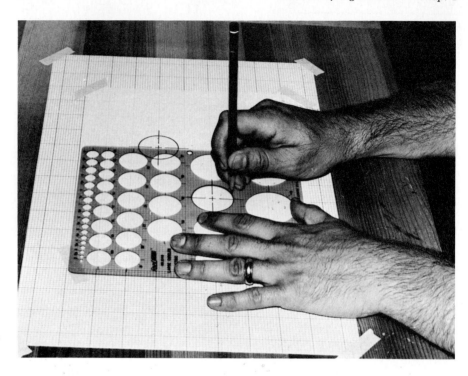

**FIGURE 1-4** Using a circle template. Remember to keep the pencil vertical.

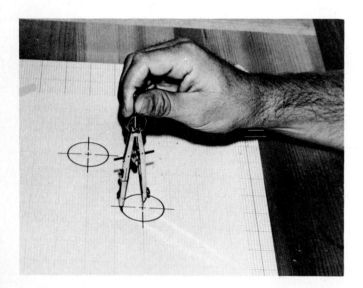

**FIGURE 1-5** Using a compass.

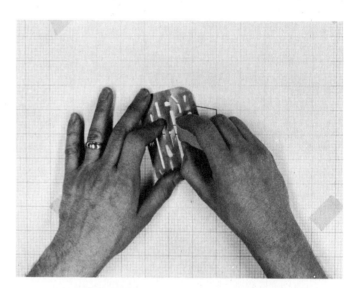

**FIGURE 1-6** Using an erasing shield.

holder across the paper, as pushing can result in torn paper. Figure 1-4 shows how to use a circle template, Figure 1-5 shows how to use a compass, and Figure 1-6 shows how to use an erasing shield.

## 1-3 LINES

Many types of lines are used on technical drawings: visible, hidden, center, leader, and phantom, to name but a few. Figure 1-7 illustrates several types. Lines used for dimensioning are illustrated in Figure 1-8.

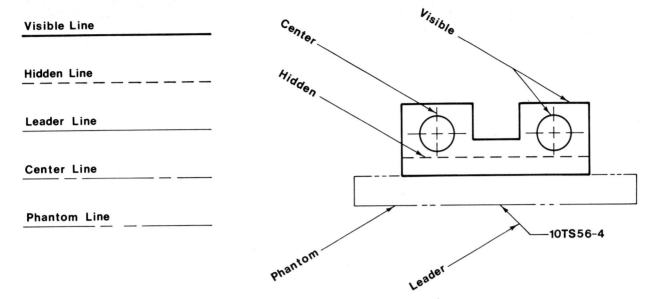

**FIGURE 1-7**  Different kinds of lines used on technical drawings.

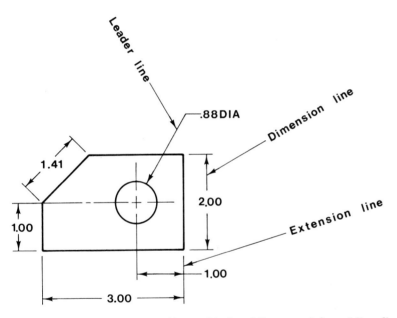

**FIGURE 1-8**  Different kinds of lines used for adding dimensions to a technical drawing.

## 1-4  LETTERING

Figures 1-9 and 1-10 show the shape and style of the letters and numbers most often used on technical drawings. Either the vertical or inclined style is acceptable, although the two styles should never be mixed. The most widely accepted height for letters and numbers is 1/8 (0.13) inch to 3/16 (0.19) inch.

To help ensure even lettering, drafters use guidelines as shown in Figure 1-11. It is acceptable to leave guidelines on the drawing—they need not be erased after the lettering has been completed.

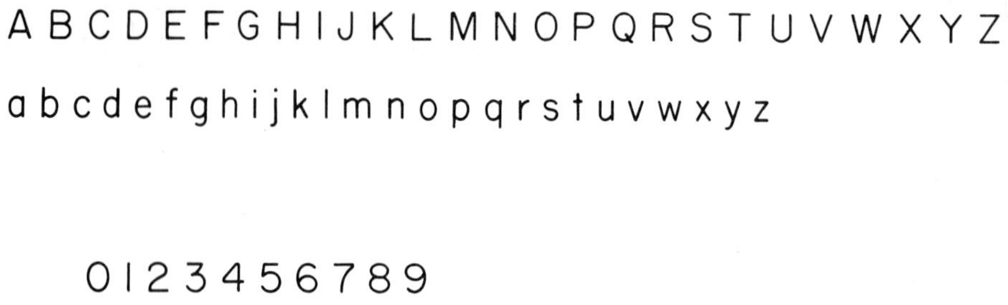

A B C D E F G H I J K L M N O P Q R S T U V W X Y Z
a b c d e f g h i j k l m n o p q r s t u v w x y z
0 1 2 3 4 5 6 7 8 9

**FIGURE 1-9** Vertical-style letters.

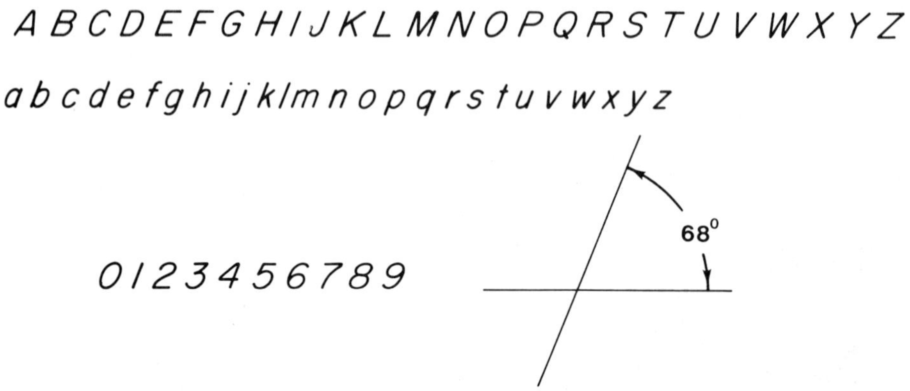

A B C D E F G H I J K L M N O P Q R S T U V W X Y Z
a b c d e f g h i j k l m n o p q r s t u v w x y z
0 1 2 3 4 5 6 7 8 9
68°

**FIGURE 1-10** Inclined-style letters.

1/8
USE ONLY CAPITOL LETTERS

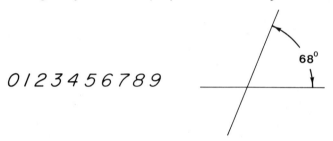

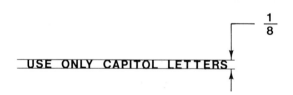

**FIGURE 1-11** Use guidelines (very light horizontal lines) to help ensure even lettering.

## 1-5 GEOMETRIC CONSTRUCTIONS

This section includes six geometric constructions which are often required when creating electronic drawings. They are presented in step-by-step format in Figures 1-12 through 1-17.

## 1-6 ORTHOGRAPHIC VIEWS

*Orthographic views* are two-dimensional drawings which are used to define three-dimensional shapes. They are used in lieu of three-dimensional pictures because they permit much greater drawing accuracy.

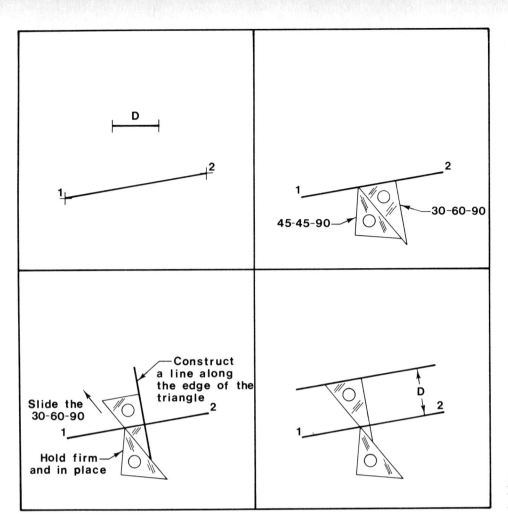

FIGURE 1-12 How to draw parallel lines at a distance $D$ using two triangles.

FIGURE 1-13 How to bisect an angle.

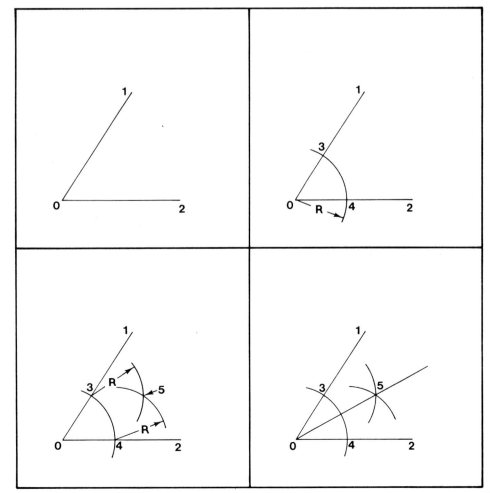

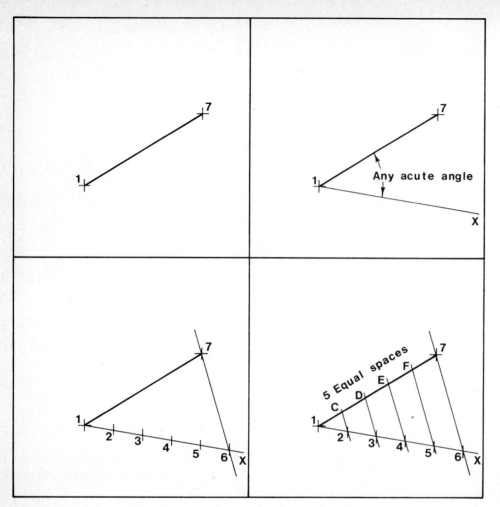

**FIGURE 1-14** How to divide a line into any number of equal spaces.

**FIGURE 1-15** How to draw a fillet of radius $R$ to an obtuse angle.

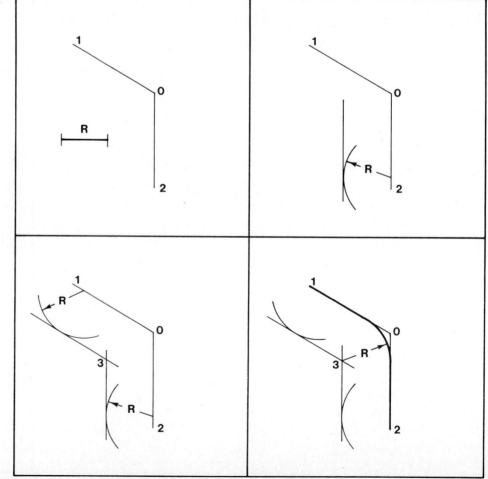

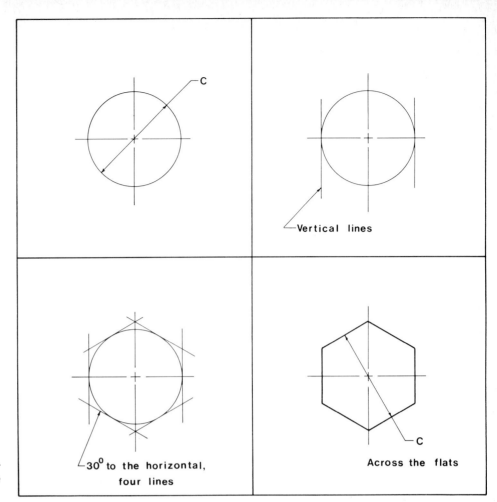

**FIGURE 1-16** How to draw a hexagon for a given distance across the flats.

C

Vertical lines

30° to the horizontal, four lines

Across the flats

**FIGURE 1-17** How to draw an ellipse for a given major and minor axis.

MINOR AXIS

MAJOR AXIS

X

A — O — B

Y

XC

C

X

OA

D

A — O — B

Y

DE=EB

X

D

E

O

A — B

G

H

Y

HX

X

G

A — O — G — B

GB

GB

H

Y

HX

Erase excess lines

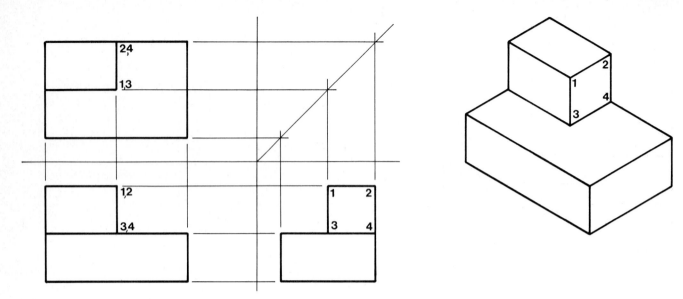

**FIGURE 1-18**  Object  and  its  front,  top,  and  right-side orthographic views.

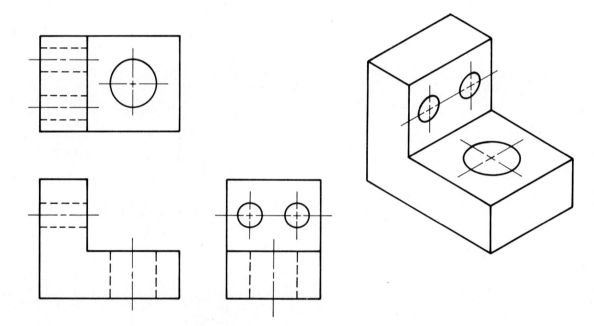

**FIGURE 1-19**  Object  and  its  orthographic  views,  which include hidden lines.

Figure 1-18 illustrates an object together with the front, top, and right-side orthographic views of the object.

*Hidden lines* are used to picture surfaces and edges which are not directly visible in orthographic views. Figure 1-19 pictures an object whose orthographic views contain hidden lines.

Inclined surfaces are drawn orthographically as shown in Figure 1-20, and rounded surfaces are drawn as shown in Figure 1-21. In each case only the profile orthographic views show the true shape of the object.

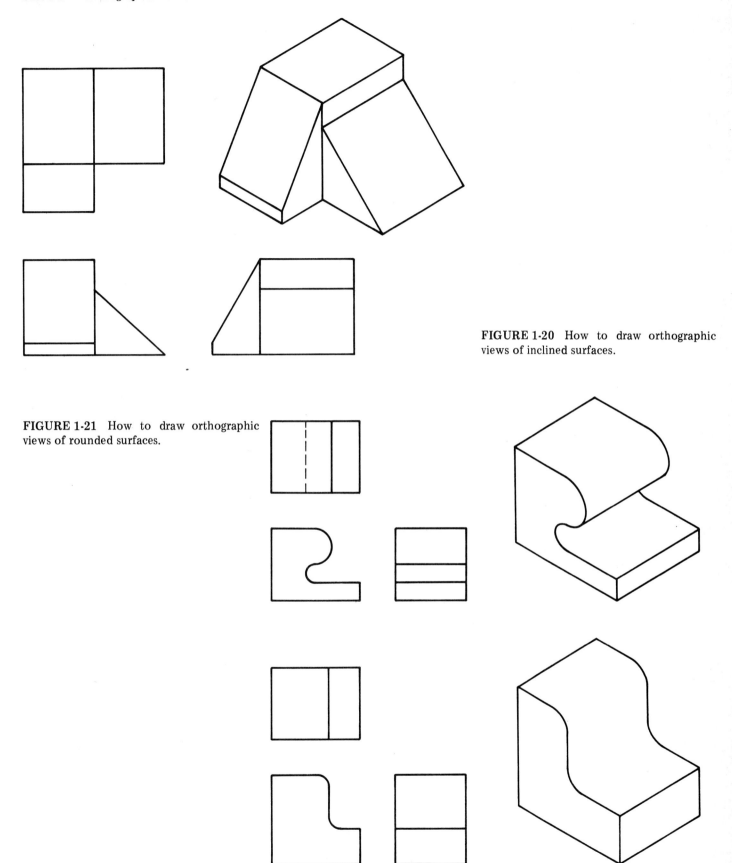

**FIGURE 1-20** How to draw orthographic views of inclined surfaces.

**FIGURE 1-21** How to draw orthographic views of rounded surfaces.

## 1-7   SECTIONAL VIEWS

*Sectional views* are used to expose the internal surfaces of an object that would otherwise be hidden from direct view. Figure 1-22 illustrates a comparison between orthographic views and sectional views. *Cutting-plane lines* (see Figure 1-23) are used to define where the sectional view is to be taken, and *section lines* (see Figure 1-24) are used to designate those surfaces which have been cut when creating a sectional view. If

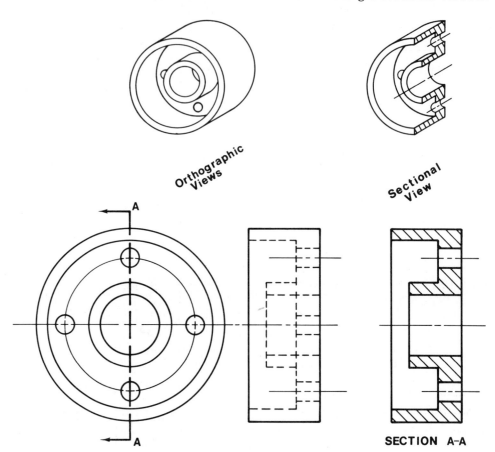

**FIGURE 1-22**   Orthographic  and  sectional  views  of  the same object.

**FIGURE 1-23**   How to draw a cutting-plane line.

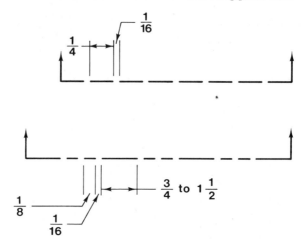

more than one sectional view is taken on the same object, the sectional views are aligned as shown in Figure 1-25 and identified by letters.

A *broken-out sectional view* is used when less than a complete sectional view is desired. It is like peeling back part of the outside surface so as to expose part of the internal surfaces. Figure 1-26 illustrates a broken-out sectional view.

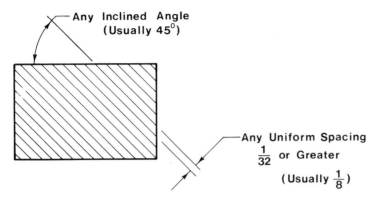

FIGURE 1-24   Section lines.

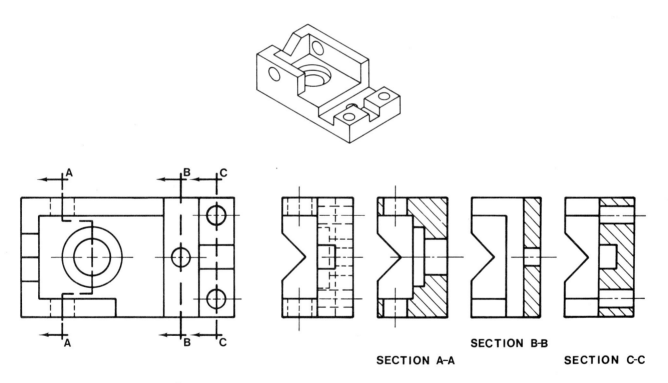

FIGURE 1-25   Multiple sectional views of the same object.

FIGURE 1-26   Broken-out sectional view.

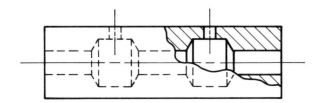

## 1-8  BASIC DIMENSIONING

Dimensions are used to define the size of an object. They are added to a drawing by using extension, dimension, and leader lines as illustrated in Figure 1-8.

All lettering used when dimensioning should be at least ⅛ (0.13) inch high. Lettering smaller than ⅛ (0.13) inch is difficult to read, especially on blueprints, and is likely to cause manufacturing errors.

Dimensions may be placed on a drawing using either the *unidirectional* or *aligned* systems, although the unidirectional is preferred. Figures 1-27, 1-28, and 1-29 illustrate how to dimension holes, arcs, angles, and small distances. All examples have been dimensioned using the unidirectional system.

**FIGURE 1-27**  Examples of unidirectional dimensioning system.

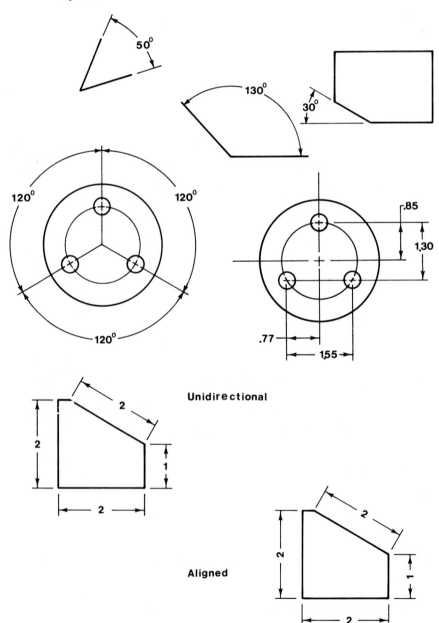

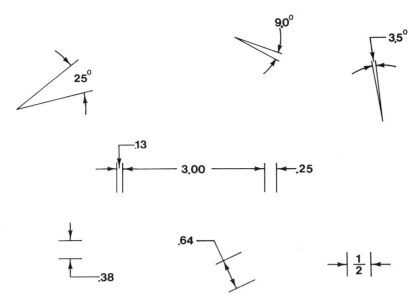

**FIGURE 1-28**  How to dimension small angles and distances.

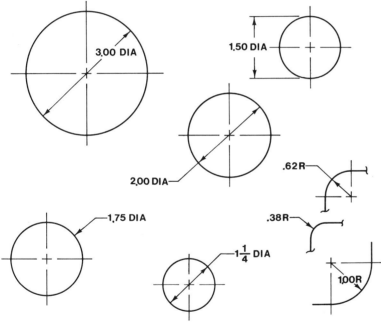

**FIGURE 1-29**  How to dimension holes and arcs.

## 1-9   METRICS

The metric system is based on a fixed unit of distance, the meter. A meter is divided into smaller units, either centimeters or millimeters. There are 100 centimeters or 1000 millimeters to a meter.

The abbreviation for a millimeter is "mm" (5 mm, 26 mm, etc.). Figure 1-30 shows a millimeter scale together with a few sample measurements.

For your convenience, conversion tables for inches to millimeters and millimeters to inches have been included on the inside back cover of this book.

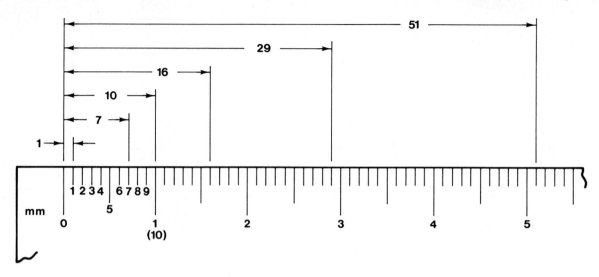

**FIGURE 1-30**  Millimeter scale together with some sample measurements.

**Simplfied**                                                    **Schematic**

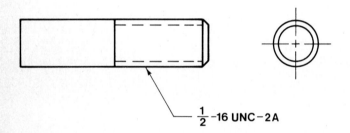

$\frac{1}{2}$-16 UNC-2A

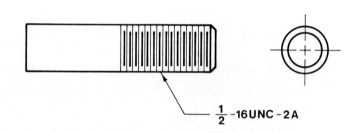

$\frac{1}{2}$-16UNC-2A

**Detailed**

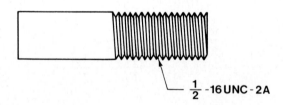

$\frac{1}{2}$-16UNC-2A

**FIGURE 1-31**  Simplified, schematic, and detailed representations of a screw thread.

**FIGURE 1-32**  Meaning of a standard thread notation.

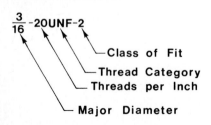

$\frac{3}{16}$-20UNF-2

Class of Fit
Thread Category
Threads per Inch
Major Diameter

## 1-10   FASTENERS

There are three different ways to present threaded fasteners on a drawing: the *simplified*, *schematic*, and *detailed* representations. Figure 1-31 shows the three types of representations.

Regardless of which representation is drawn, the same thread notations are used. Figure 1-32 defines the meaning of a standard thread notation.

When drawing threaded holes, always account for the pilot drill hole (a small hole drilled before the threads are cut). When showing a

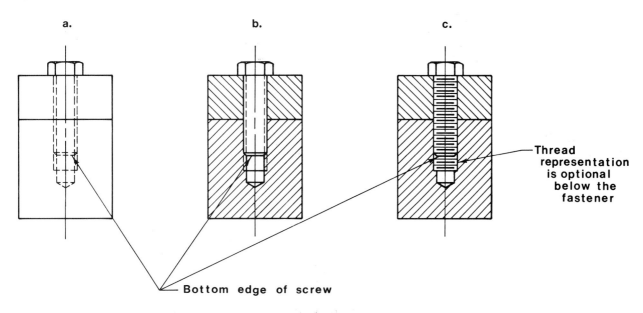

**FIGURE 1-33**   How to draw a fastener in a threaded hole.

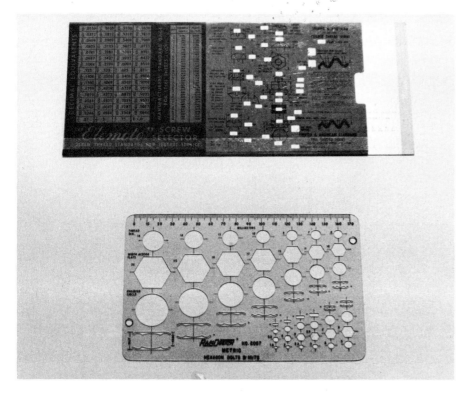

**FIGURE 1-34**   Template that is helpful when drawing
bolts and nuts, and an Elemoto screw selector.

fastener inserted in a threaded hole, remember to draw clearly the
bottom edge of the fastener, the remaining unused threaded portion of
the hole, and the pilot drill hole, as shown in Figure 1-33.

Drafters often use fastener templates as guides when drawing
fasteners. Figure 1-34 illustrates one of the many fastener templates
commercially available, and an Elemoto screw selector, which is helpful
in picking thread and fastener sizes.

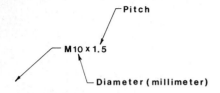

**FIGURE 1-35** Meaning of a metric thread notation.

Metric threads are designated on drawings as shown in Figure 1-35. An "M" indicates a metric thread. The first number specifies the major diameter of the thread, and the second number the pitch or the number of threads per inch. Sometimes the pitch number is omitted. This omission implies the use of a standard coarse thread of the diameter designated.

## 1-11   ASSEMBLY DRAWINGS

Mechanical drawings are generally classified as either assembly drawings or detail drawings. The *assembly drawing* shows all the pieces of a device in their assembled positions. *Detail drawings* show individual parts and should include all information (dimensions, special considerations, etc.) necessary to make the part.

To help develop an understanding of the relationship between assembly drawings and detail drawings, remember:

*Assembly drawings show you how to put the parts together.*

*Detail drawings show you how to make the parts.*

**FIGURE 1-36**   Exploded drawing of a clamping device.

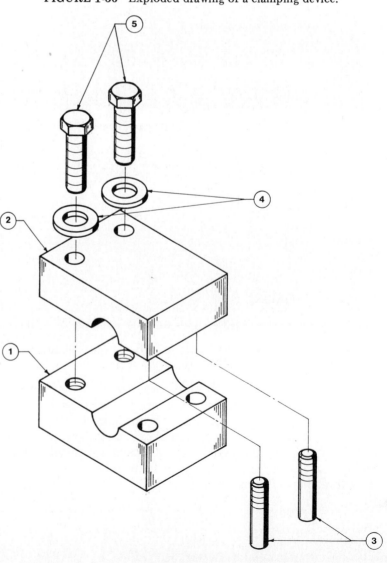

Consider the clamping device shown in Figure 1-36. To define this device properly, we need to prepare an assembly drawing, three detail drawings, and a parts list.

Figure 1-37 shows the assembly drawing for the device shown in Figure 1-36. Note that the assembly drawing contains *no hidden lines.* The two guide pins are shown in partial sectional views, as they would otherwise not be seen. Parts, such as the guide pins, which are internal assembly parts are not shown using hidden lines but are always shown using sectional views. If the internal parts are small and difficult to show clearly, the sectional view may be redrawn next to the assembly view but using an increased scale.

Each part on the assembly drawing is assigned a number. The numbers are circled. Some companies require complete part numbers (usually very long), but most permit assembly drawing numbers which are used only on the specific assembly drawing and are referenced to the parts list. The assembly drawing itself is numbered 43S100.

Figure 1-38 shows the parts list that would accompany the assembly drawing shown in Figure 1-37. There are many different styles of parts lists, and Figure 1-38 is only a representative sample.

The left-hand column of the parts list lists the numbers as defined on the assembly drawing. Part number 1 on the assembly drawing is a base clamp, as stated in the description column, and is to be made from SAE1020 steel, as stated in the material column.

Materials are specified using standardized codes such as SAE1020,

**FIGURE 1-37**   Assembly drawing of a clamping device.

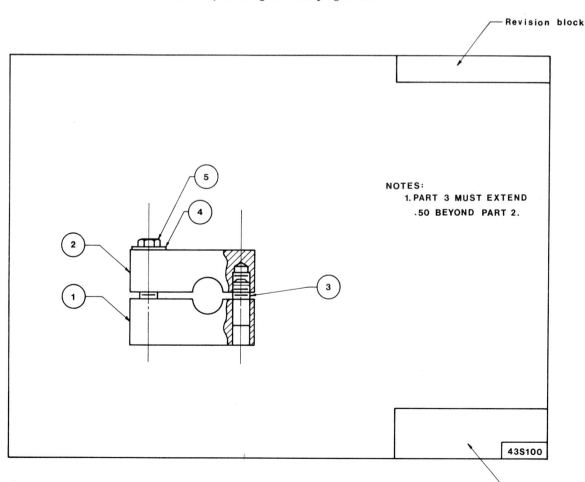

| NO | DESCRIPTION | MATL | PART NUMBER | QTY |
|----|-------------|------|-------------|-----|
|  |  |  |  |  |
|  |  |  |  |  |
| 5 | SCREW | ST | .31-18 UNC × 1.38 HEX HEAD | 2 |
| 4 | WASHER | ST | .375 × .875 × .083 | 2 |
| 3 | GUIDE PIN | SAE 1020 | 43S103 | 2 |
| 2 | TOP CLAMP | SAE 1020 | 43S102 | 1 |
| 1 | BASE CLAMP | SAE 1020 | 43S101 | 1 |
| NO | DESCRIPTION | MATL | PART NUMBER | QTY |

FIGURE 1-38   Sample parts list.

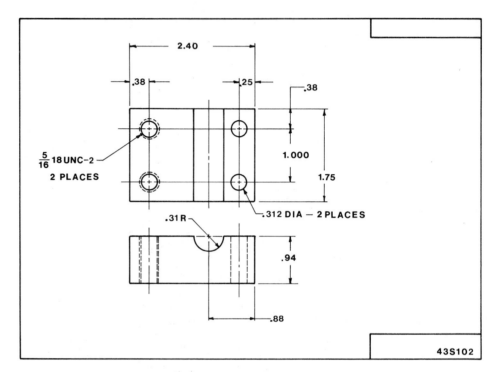

FIGURE 1-39   Detail drawing of a base clamp.

the Society of Automotive Engineers' listing 1020 for steels. There are many other materials codes in use.

The number 43S101 is the part number assigned to the base clamp and is also the drawing number of the base clamp detail drawing. Do not confuse this part number with the circled number used on the assembly drawing. The smaller numbers are for the assembly drawing *only*; the larger number is the official company-assigned part number and is the one used in all references and correspondence. The right-hand column states the number of parts required for assembly.

Standard parts are parts that can be purchased from manufacturers and used directly on the assembly. No modifications are required for standard parts. They are used in the "as received" condition. For example, the screw, identified as number 5 on the assembly drawing, is a standard part. A company would purchase it rather than make it themselves. The abbreviation ST for steel is sufficient to define the material unless a specific type of steel is required. No part number and no detail drawing are needed. The size can be defined completely using the American National Standards Institute (ANSI) code (see Section

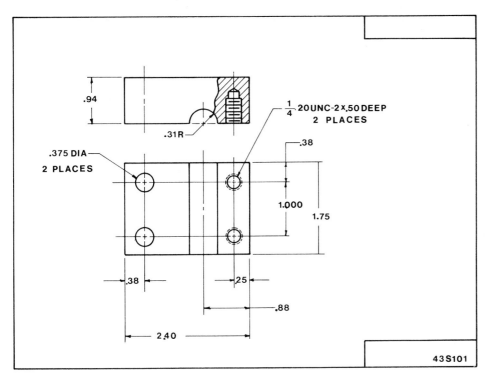

FIGURE 1-40   Detail drawing of a top clamp.

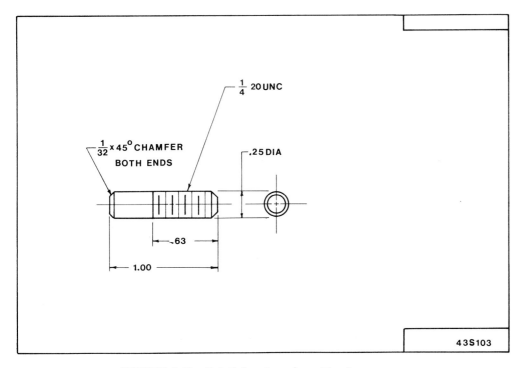

FIGURE 1-41   Detail drawing of a guide pin.

1-10). Washer sizes are also standardized and are listed by the inside diameter, outside diameter, and thickness.

Note that the parts list is numbered from the bottom up, and that the top has additional spaces for more parts. This is done so that future changes can easily be added to the list.

Figures 1-39, 1-40, and 1-41 are detail drawings that would supplement the assembly drawing. A detail drawing should contain all information necessary to manufacture a part.

## PROBLEMS

1-1    Draw a line that slopes up from left to right at 27° to the horizontal and is 2.50 inches long. Draw two more lines parallel to the first, one of which is 1.00 inch above the first, and the other, 0.75 inch below.

1-2    Draw a 45° angle and bisect it.

1-3    Draw a line that slopes down from left to right at 30° to the horizontal and is 3.25 inches long. Divide the line into five equal parts.

1-4    Draw an angle of 115° and draw a round 0.38 inch in diameter within the angle.

1-5    Draw a hexagon of 1.00 inch across the flats.

1-6    Draw a hexagon of 0.63 inch across the flats.

1-7    Draw an ellipse whose minor axis is 1.25 inches and whose major axis is 2.88 inches.

1-8 and 1-9    Redraw Figures P1-8 and P1-9, including the dimensions.

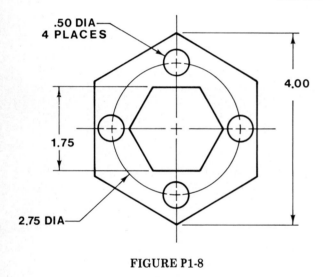

FIGURE P1-8

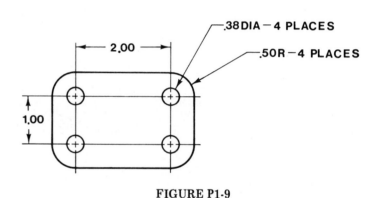

FIGURE P1-9

1-10    Redraw Figure P1-10, including the dimensions. The dimensions are in millimeters.

FIGURE P1-10

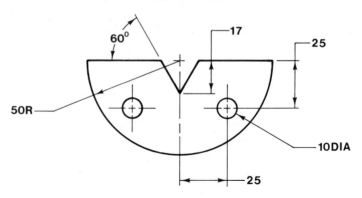

**1-11**
**and 1-12**    Redraw Figures P1-11 and P1-12.

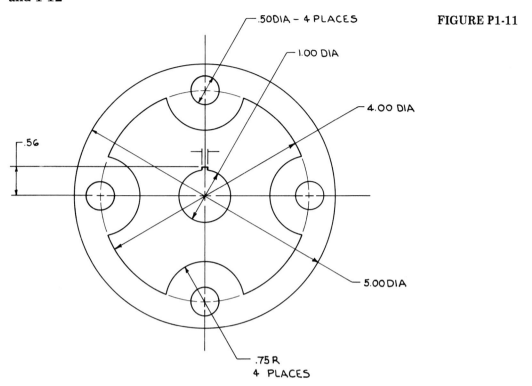

**FIGURE P1-11**

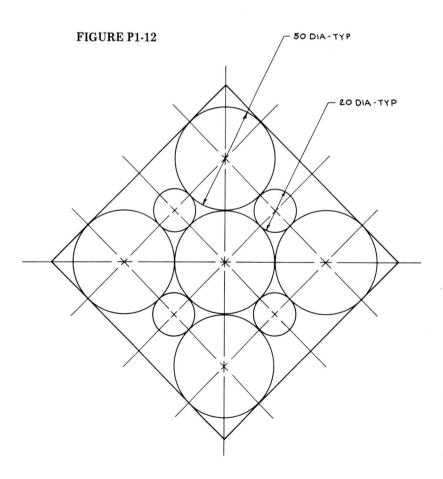

**FIGURE P1-12**

**1-13**    Redraw Figure P1-13. All lines must be of exactly the same thickness and density.

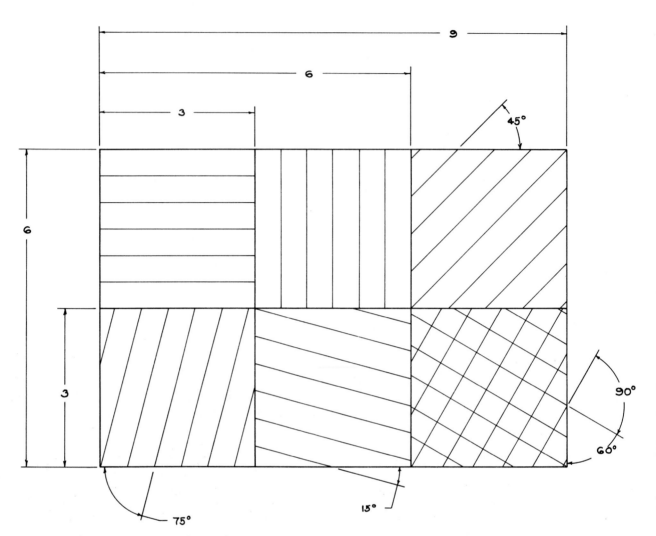

**FIGURE P1-13**

**1-14**    Redraw Figure P1-14, including the chart. After drawing the figure, measure the figure and add the appropriate values to the chart.

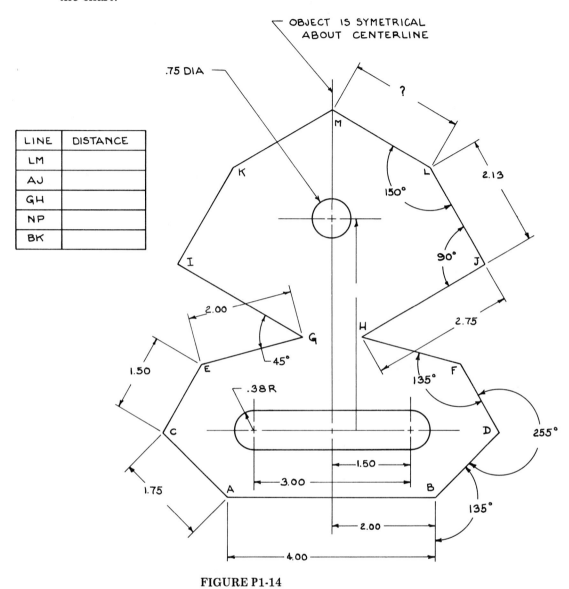

| LINE | DISTANCE |
|------|----------|
| LM   |          |
| AJ   |          |
| GH   |          |
| NP   |          |
| BK   |          |

FIGURE P1-14

**1-15 and 1-16**  Redraw Figures P1-15 and P1-16 and add all dimensions necessary. The figures are drawn to scale and may be measured directly.

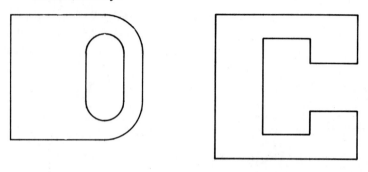

FIGURE P1-15                    FIGURE P1-16

**1-17 through 1-23**  Draw the front, top, and right-side views of Figures P1-17 through P1-23.

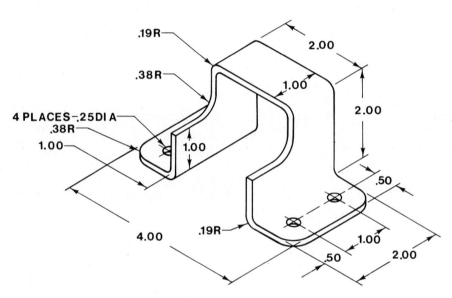

FIGURE P1-17

FIGURE P1-18

MATL

$\frac{1}{8}$ 2024 – T4 AL

FIGURE P1-19

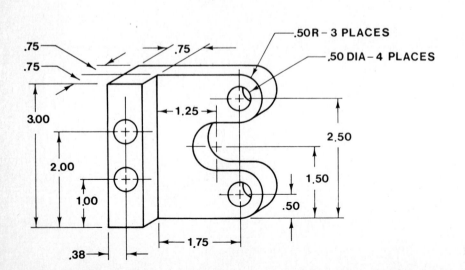

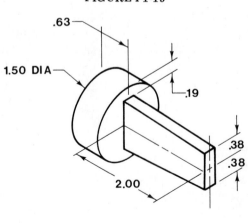

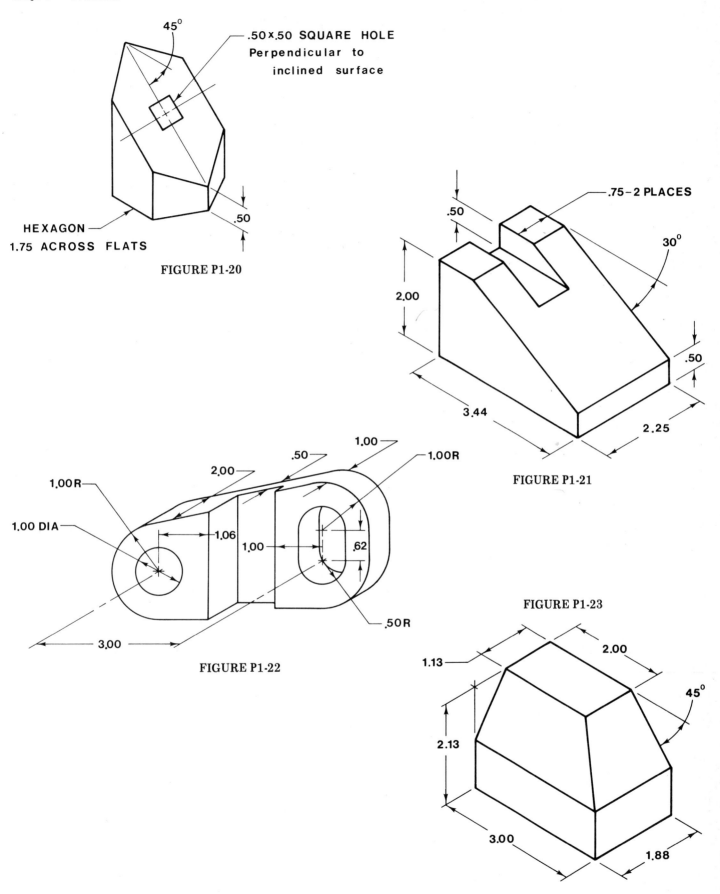

45°

.50×.50 SQUARE HOLE
Perpendicular to
inclined surface

HEXAGON
1.75 ACROSS FLATS

.50

FIGURE P1-20

.50

2.00

.75–2 PLACES

30°

.50

3.44

2.25

FIGURE P1-21

1.00R

1.00R

1.00 DIA

2.00

.50

1.00

1.00R

1.06

1.00

.62

.50R

3.00

FIGURE P1-22

FIGURE P1-23

1.13

2.00

45°

2.13

3.00

1.88

**1-24**   Redraw the front and right-side views given in Figure P1-24 and add the top view.

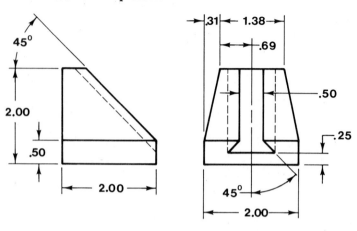

FIGURE P1-24

**1-25**   Redraw the front and top views of Figure P1-25 and add the right-side view.

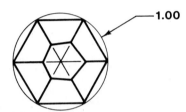

**1-26 through 1-32**   Draw three views each of Figures P1-26 through P1-32.

FIGURE P1-26

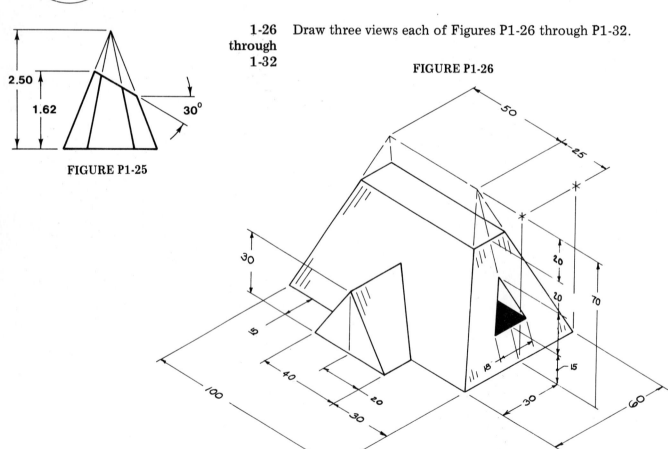

FIGURE P1-25

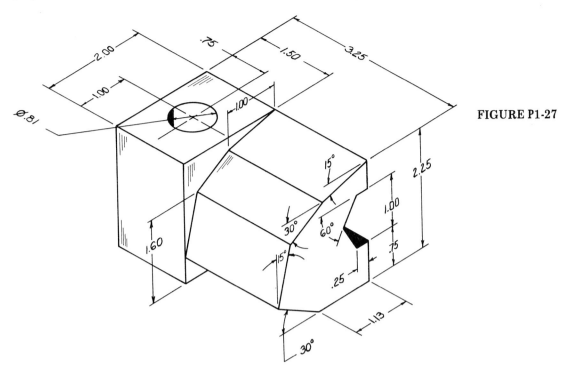

**FIGURE P1-27**

**FIGURE P1-28**

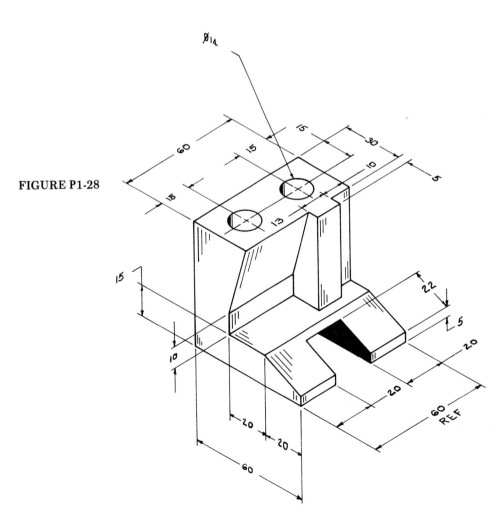

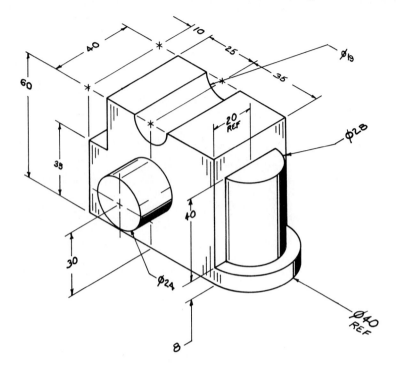

**FIGURE P1-29**

**FIGURE P1-30**

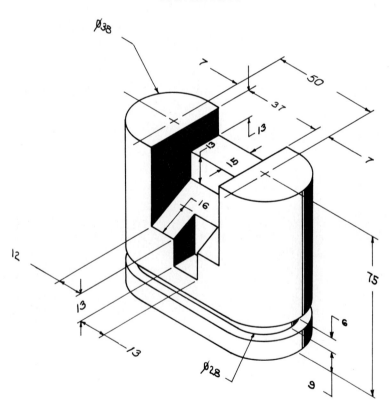

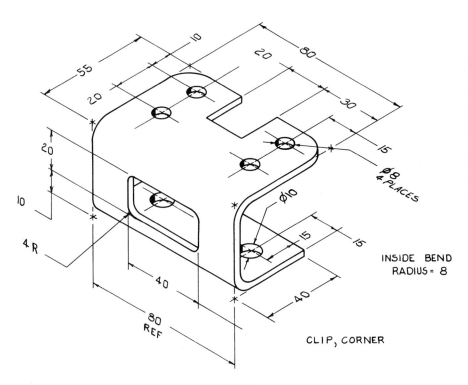

CLIP, CORNER

INSIDE BEND
RADIUS = 8

**FIGURE P1-31**

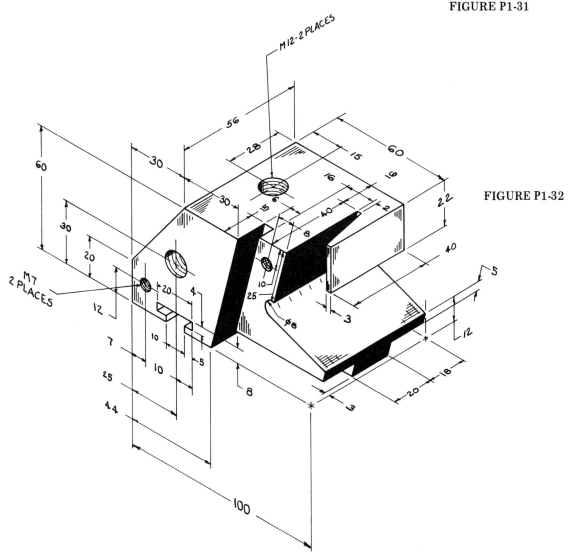

**FIGURE P1-32**

1-33    Redraw Figure P1-33 and replace the front view with a sectional view.

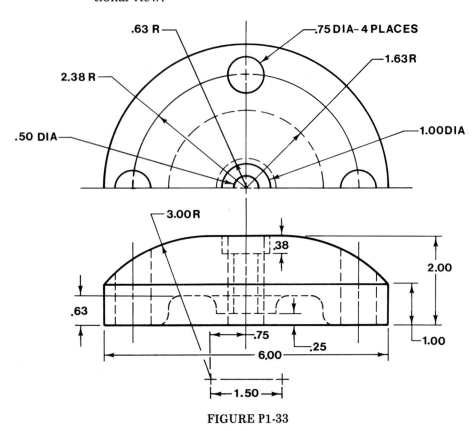

FIGURE P1-33

1-34    Redraw Figure P1-34 and fasteners.
   a.  ¼-20 UNC × 1.25 hex head
   b.  ⅜-24 UNF × 1.50 square head
   c.  ½-13 UNC × 1.00 headless stud
   d.  ⁷⁄₁₆-20 UNF × 1.44 hex head

FIGURE P1-34

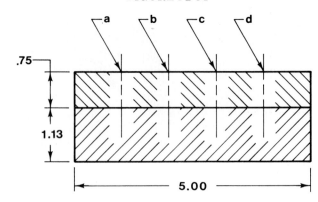

1-35   From the information given in Figure P1-35(a) and (b), pre-
       pare an orthographic assembly drawing and appropriate
       parts list.

**FIGURE P1-35**

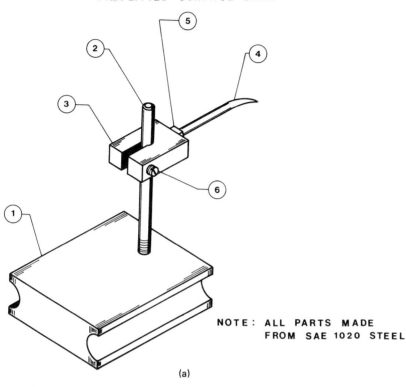

SIMPLIFIED SURFACE GAGE

NOTE: ALL PARTS MADE
FROM SAE 1020 STEEL

(a)

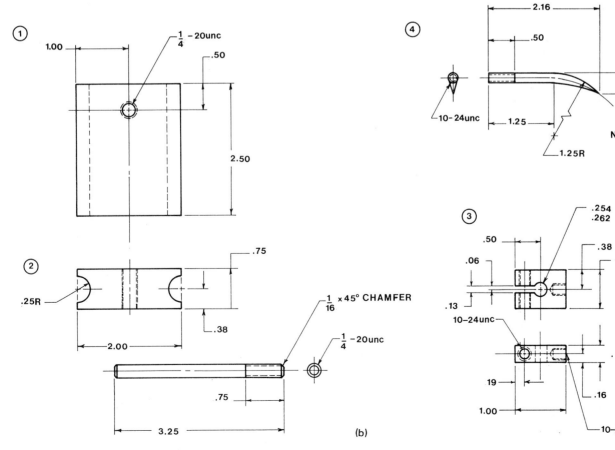

(b)

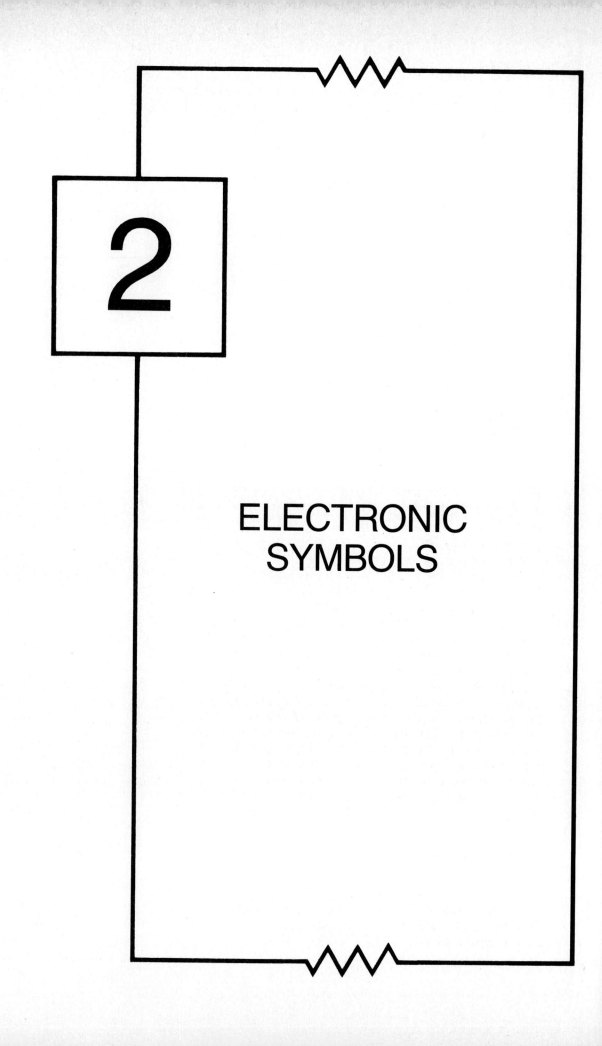

# 2

# ELECTRONIC
# SYMBOLS

## 2-1 INTRODUCTION

In this chapter we present some of the many symbols used to represent components in electronic drafting. Also included are discussions on how to use an electronic symbol template and how to create your own symbols.

So that you can better understand the sizes and proportions of each symbol, the symbols have been drawn twice their usual size. It should be pointed out that the sizes given in this chapter, although in agreement with established national standards, are intended as guides, not as rigorous, absolute measurements. They may be varied slightly as necessary, as long as the proportions of the symbol remain the same.

Consider, for example, the symbol for a resistor shown in Figure 2-1. The one marked "per standards" is drawn exactly per MIL-STD-15 and ANSI Y32.2 specifications. In addition, two correct and two incorrect variations are shown. In each case, the acceptable variations are correct because they increase the size of the symbol but do not change the proportions, whereas the unacceptable variations do vary the proportions. When preparing your drawings, choose a scale that best (most clearly) presents the symbols in a neat, easy-to-follow size.

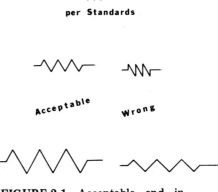

FIGURE 2-1 Acceptable and incorrect ways to draw electronic symbols.

## 2-2 HOW TO USE A SYMBOL TEMPLATE

Most drafters use symbol templates as guides when drawing electronic symbols. They enable drafters to work much faster than they would were they using standard drafting equipment (T-square, triangle, compass, etc.) to create the symbols.

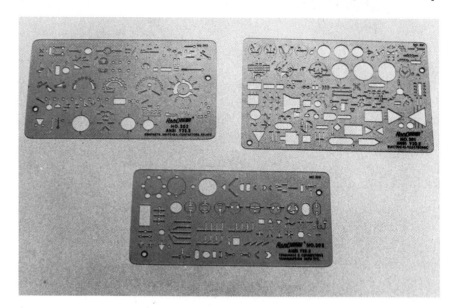

FIGURE 2-2   Electronic symbol templates.

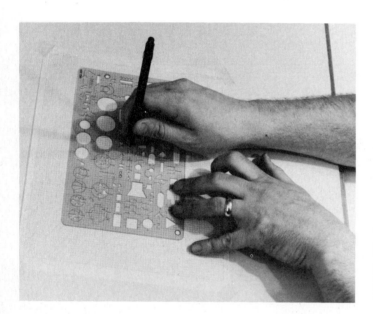

FIGURE 2-3   How to use a symbol template.

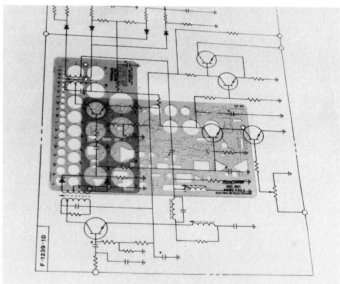

FIGURE 2-4   If template cutouts are too wide, slide another template underneath and draw as shown in Figure 2-3.

Many different types of symbol templates are available commercially; Figure 2-2 shows a few. Some are for general use, others are for more specific purposes, such as switches or logic symbols. Regardless of which you choose, they should conform in size with either ANSI Y32.2 or MIL-STD-15 standards.

To use a symbol template, align the appropriate symbol over the place where you want it to appear on the drawing and then, holding the pencil as nearly vertical as possible (see Figure 2-3), trace the symbol using the template as a guide. If a double line appears, it means that the hole cut in the template is too wide. This can be overcome by sliding another template or a triangle under the symbol, as shown in Figure 2-4. This will raise the template to a thicker portion of the tapered drawing lead. If the lead will not fit into the template, resharpen the lead to a thinner taper so that it will fit.

It should be pointed out that some templates are manufactured in such a way that it is impossible to draw the symbols in one alignment. In these instances, the symbols must be drawn in parts, using several different cutouts on the template. When this is the case, the final symbol should appear neat and clear, just as though it were drawn in one smooth motion.

## 2-3   DRAWING PAPER

Most electronic drafters prepare their drawings on grid paper, such as that shown in Figure 2-5. The grid is very helpful in drawing wiring paths and aligning symbols.

The grid is usually printed in nonreproducible blue so that it will not appear on the blueprints made from it. The grid is available in different sizes—$\frac{1}{8}$, $\frac{1}{10}$, $\frac{1}{5}$, $\frac{1}{4}$— and in metric sizes, but most drafters use the $\frac{1}{10}$ or $\frac{1}{8}$ size.

**FIGURE 2-5** Sheet of drawing paper with a nonreproducible grid. Squares are eight to the inch.

## 2-4  INTEGRATED CIRCUITS

Integrated circuits (ICs) are complete circuits mounted on a silicon chip. They are connected to a large circuit through pins mounted along the edges of the silicon chip.

The graphic symbol for an IC is a rectangle, as shown in Figure 2-6. The IC device shown in Figure 2-6 is a low-power operational amplifier sold by Radio Shack. The number ICL7611 is the chip's part number. Different manufacturers and distributors use different numbers, but each publishes a replacement guide or equivalent listing to permit easy interchangeability of equivalent devices.

The pins of an IC device are numbered consecutively starting with 1. Pins are always located along the longer sides of the rectangle. If the IC is right-side up, pin 1 is always the upper left pin. A notch or printed semicircle is manufactured on the IC to identify its correct orientation. The notch or printed semicircle is not always included in the graphic representation of the IC.

An amplifier (amp) is a type of IC device which is used so often that it has its own symbol, an equilateral triangle. Figure 2-7 shows an amp symbol. Remember that an equilateral triangle has three 60° angles and three equal-length sides. Pin numbers or amps vary as to specific functions.

Amps are often used individually or in groups as part of an IC. The actual chip is still rectangular in shape, and is represented symbolically by a rectangle, but the rectangle also includes the amp symbols. Figure 2-8 shows the symbol for such an IC.

Logic symbols (see Chapter 5) are often included within the rectangular IC symbol, to help define the IC's function. Figure 2-9 shows two examples of IC symbols that include logic symbols.

Section 5-4 discusses the combining of IC and logic symbols. Section 5-3 shows how IC symbols are used in schematic diagrams.

FIGURE 2-6  Graphic symbol for an integrated circuit.

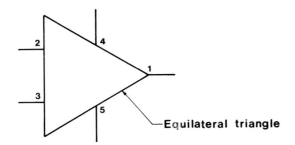

FIGURE 2-7   Graphic symbol for an amp.

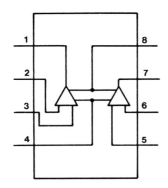

FIGURE 2-8   IC symbol that
includes amp functions.

FIGURE 2-9   IC with logic symbols.

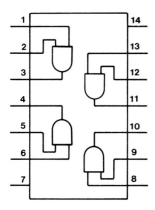

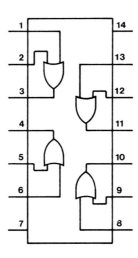

2-5   ANTENNAS

## Drawing   Specifications
## (Twice  Size)

## Per   Standards

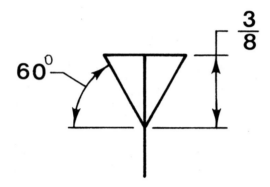

## Pictorial   Representation

## Related   Symbols

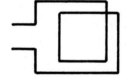

Dipole                            Loop                            Counterpoise

## 2-6   BATTERIES

**Drawing   Specifications   (Twice   Size)**          **Per   Standards**

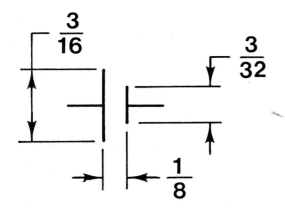

**Pictorial   Representation**

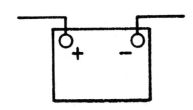

**Related   Symbols**

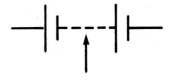

Multicell                    With Taps                    Adjustable

## 2-7   CAPACITORS

**Drawing   Specifications          Per   Standards**
**(Twice   Size)**

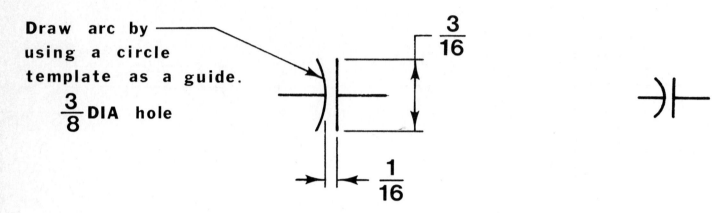

**Draw   arc   by**
**using   a   circle**
**template   as   a   guide.**

$\frac{3}{8}$ **DIA   hole**

$\frac{3}{16}$

$\frac{1}{16}$

**Pictorial   Representation**

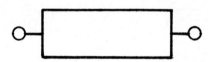

**Related   Symbols**

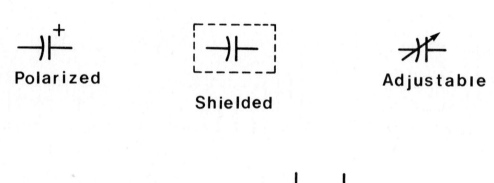

Polarized          Shielded          Adjustable

Split-Stator          Dual

## 2-8  DIODES

**Drawing  Specifications**          **Per  Standards**
**(Twice  Size)**

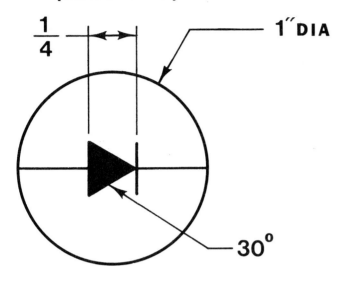

**Pictorial   Representation**

**Related   Symbols**

**Capacitive**       **Photosensitive**      **Temperature**
                                              **Dependent**

**Unidirectional**        **Tunnel**            **Zener**

## 2-9   GROUNDS

**Drawing Specifications (Twice Size)**       **Per Standards**

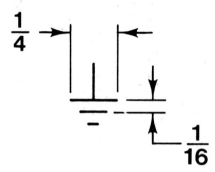 

**Pictorial Representation**

**Related Symbols**

**Chassis**

## 2-10   INDUCTORS

### Drawing Specifications
### (Twice Size)

### Per Standards

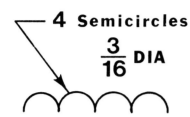

## Pictorial   Representation

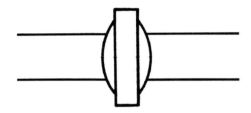

## Related   Symbols

**Magnetic
Core**

**Tapped**

**Adjustable**

**Continuous
Adjustable**

## 2-11   INTEGRATED CIRCUITS

## Drawing  Specifications                              Per  Standards

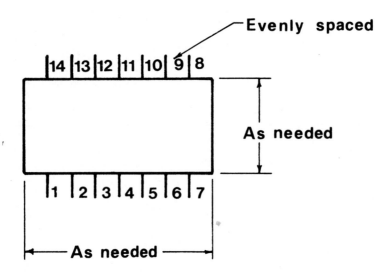

### Pictorial  Representation

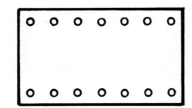

### Related  Symbols

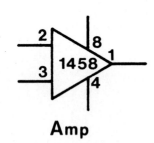

Amp

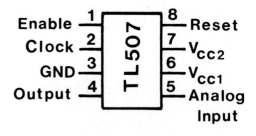

Converter

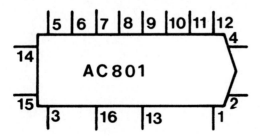

8-Bit  digital  to  analog
converter

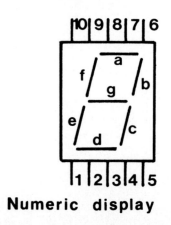

Numeric  display

## 2-12  METERS

**Drawing   Specifications          Per   Standards**
**Pictorial   Representation**

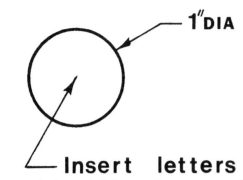

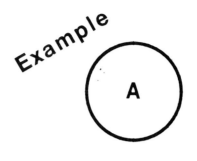

## LETTER   CODE

| | |
|---|---|
| A – Ampmeter | OHM – Ohmmeter |
| CRO – Oscilloscope | PH – Phasemeter |
| DB – Decibel Meter | $t^o$ – Temperature Meter |
| F – Frequency Meter | V – Voltmeter |
| I – Indicating | W – Wattmeter |

2-13  RESISTORS

**Drawing Specifications**
**(Twice Size)**

**Per  Standards**

**3 Peaks, both top**
**and  bottom**

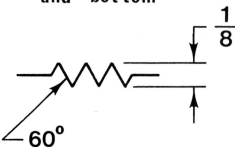

**Pictorial   Representation**

**Related   Symbols**

**Adjustable**                          **With  Taps**

## 2-14  SWITCHES

### Drawing  Specifications          Per  Standards
### (Twice Size)

**No  dimensions**
**Drawn  by  eye**

### Pictorial  Representation

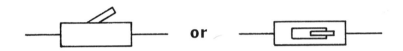

### Related  Symbols

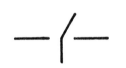

**Double-Throw**            **2 Pole  Double**            **Toggle**
                            **Throw  Switch**

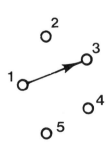

**Pushbutton**              **Multiposition**            **Rotary**

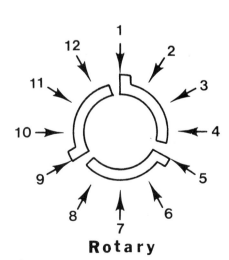

## 2-15  TRANSFORMERS

**Drawing  Specifications**
**(Twice  Size)**

**Per  Standards**

8  Semicircles

$\frac{3}{16}$ **DIA**

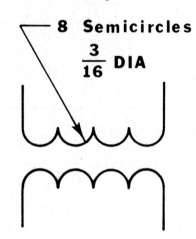

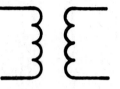

**Pictorial   Representation**

**Related   Symbols**

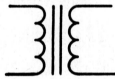

**Magnetic  Core**

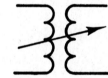

**Adjustable**

**With  Taps**

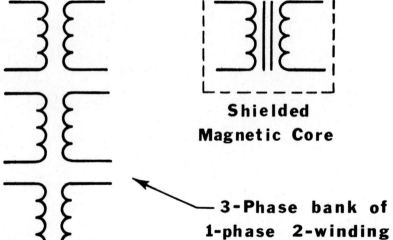

**Shielded**
**Magnetic  Core**

**3-Phase  bank  of**
**1-phase  2-winding**
**transformers**

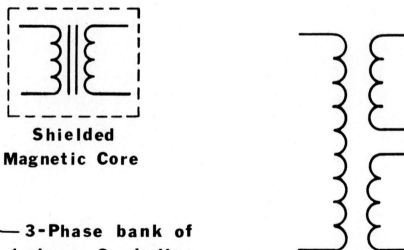

**Multiple  Taps**

## 2-16   TRANSISTORS

### Drawing Specifications (Twice Size)

### Per Standards

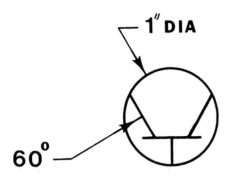

**NPN**

**PNP**

### Pictorial   Representation

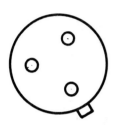

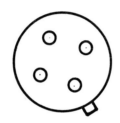

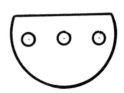

### Related   Symbols

**NPN with**
**Transverse-Biased Base**

**Unijunction**
**N-Type**

**Unijunction**
**P-Type**

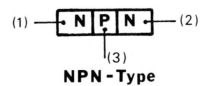

**NPN-Type**

## 2-17   OTHER SYMBOLS

**Amplifier**

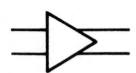

**Crystal**

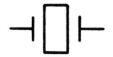

**Bell**

**Delay   Function**

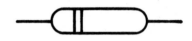

**Buzzer**

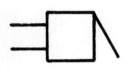

**Envelopes**

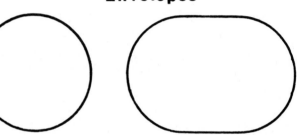

**Circuit   Breaker**

**Fuse**

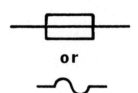

**Connectors—Power  Supplies**

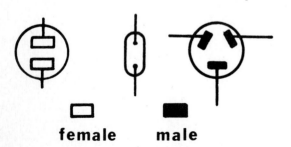

female        male

**Headset**

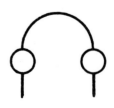

## 2-17 CONTINUED

### Lamp

### Rectifier

### Shielding

### Motor

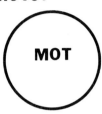

### Relay

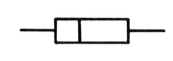

### Speaker

### Permanent Magnet

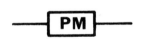

### Relay Coil

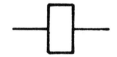

### Terminal Board

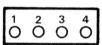

### Phase Shifter

### Repeater

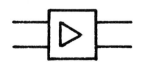

### Thermal Elements

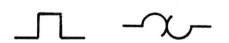

### Pickup Head

### Safety Interlock

### Thermocouple

## 2-18   SYMBOL REFERENCES

In addition to the symbols presented in this chapter, there are many more symbols used in electronic drafting. A complete listing of all standard symbols can be found in either of the following publications:

MIL-STD-15-1, Graphic Symbols for Electrical and Electronics Diagrams. Washington, D.C.: U.S. Department of Defense.

USAS Y32.2-1967, Graphic Symbols for Electrical and Electronic Diagrams. New York: American National Standards Institute.

## 2-19   CREATING YOUR OWN SYMBOLS

If, as you are preparing a drawing, you are asked to draw a component for which you are unable to find a predefined symbol, it is acceptable to create your own symbol, provided that you completely define the new symbol in the drawing's ledger. It should be emphasized that creating new symbols is acceptable *only* when you cannot find an equivalent in the symbol standards.

When creating your own symbols, any shape or lettering code may be used, although you should be careful not to use shapes or letters which are so similar to well-known symbols that the reader might mistake your new symbol for the known one. Figure 2-10 gives some hypothetical examples of new symbols.

## 2-20   OTHER WAYS TO DRAW SYMBOLS

In addition to using symbol templates as guides, there are several other ways to draw electronic symbols. If the drawing is to be done in ink, a Leroy symbol guide, shown in Figure 2-11, could be used. Dry-transfer symbols may be used in conjunction with ink, pencils, or tape drawings. Computers can be programmed to draw not only symbols but also entire drawings, including wiring paths and written data. Each of these techniques will produce clear, easy-to-read, and easy-to-reproduce drawings, but of the three described, computers offer the best possibilities because of their drawing speed and designing capabilities.

**Square Reader**

**Super Light**

**Transistor X**

**FIGURE 2-10**  Hypothetical symbols.

**FIGURE 2-11**  Using a Leroy symbol guide to draw electronic symbols.

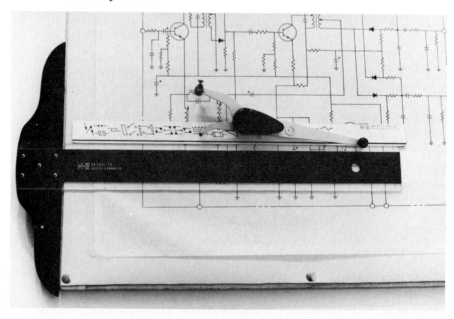

# PROBLEMS

2-1    Using the format outlined in Figure P2-1(a) and the title block as defined on the inside front cover, identify the symbols sketched in Figure P2-1(b) and letter the names next to the appropriate numbers.

**FIGURE P2-1** (a)                                            (b)

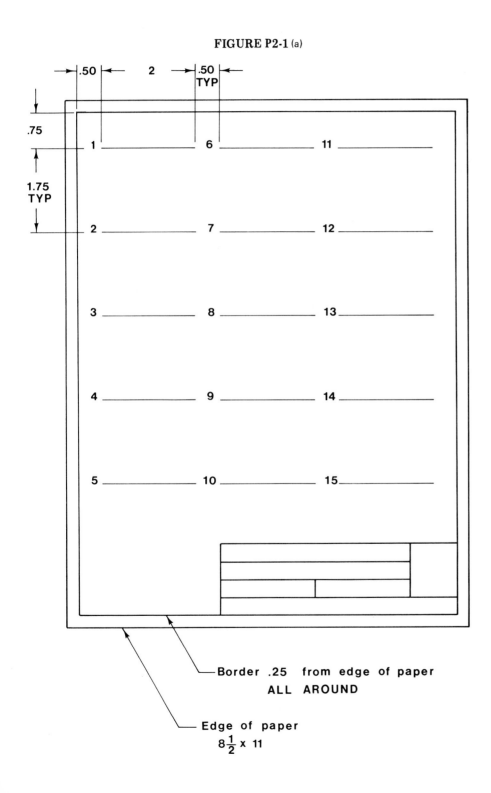

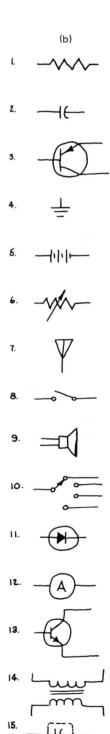

**2-2**   Using the format outlined in Figure P2-1(a), draw and label the following symbols.
1. Relay coil
2. Thermocouple
3. Fuse
4. PNP transistor
5. Dipole antenna
6. Temperature-dependent diode
7. Polarized capacitor
8. Adjustable resistor
9. Chassis ground
10. Phase shifter
11. NPN transistor
12. Tapped inductor coil
13. Battery
14. Bell
15. Circuit breaker

**2-3**   Using the format outlined in Figure P2-1(a), draw and label the following symbols.
1. Delay function
2. Loop antenna
3. Amplifier
4. Unijunction P-type transistor
5. Tapped resistor
6. Ampmeter
7. Multicell battery
8. Safety interlock
9. Counterpoise antenna
10. Ground
11. Split stator
12. Buzzer
13. Permanent magnet
14. Resistor
15. Ohmmeter

**2-4**   Using the format outlined in Figure P2-1(a), draw and label the following symbols.
1. Photosensitive diode
2. Shielded capacitor
3. Antenna
4. Continuous adjustable inductor
5. Terminal board
6. Toggle switch
7. Transformer
8. Speaker
9. Motor
10. Thermal element
11. Temperature-dependent diode
12. Voltmeter
13. Pickup head
14. Pushbutton
15. Dipole antenna

**2-5**   Draw and label the following pictorial symbols.
   1. Antenna
   2. Transistor
   3. Inductor
   4. Resistor
   5. Capacitor
   6. Battery

**2-6**   Using the format outlined in Figure P2-1(a), draw and label the following symbols.
   1. Transistor with in-line pins (use a pictorial symbol)
   2. Transistor with triangular pin location (use a pictorial symbol)
   3. General Semiconductor device 0.50 inch X 1.00 inch with five pins located on each of the 1.00-inch sides
   4. Amplifier device with five pins
   5. Transistor, N-channel type (use a pictorial symbol)
   6. Delay function
   7. General Semiconductor device 20 mm X 80 mm with four pins on each of the 80-mm sides
   8. Diode
   9. Chassis ground
   10. Ampmeter
   11. Amplifier devices with three pins
   12. Phase shifter
   13-15. A General Semiconductor device 1.50 inches X 3.50 inches. Locate 16 equally spaced pins on each of the 3.50-inch sides. Number of pins from 1 through 16 on the left side from top to bottom and 17 through 32 on the right side from top to bottom.

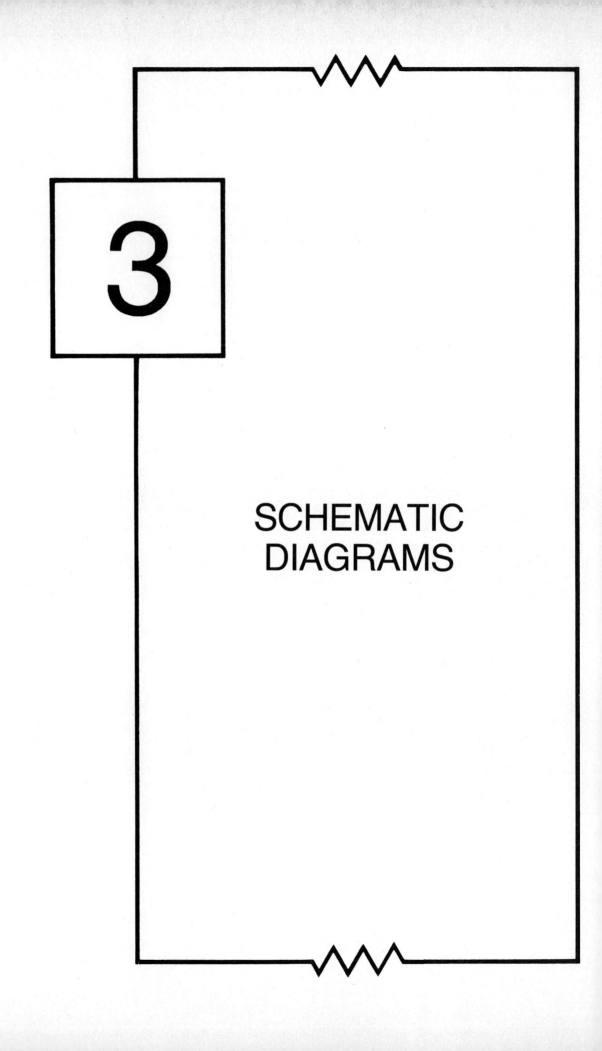

# 3

# SCHEMATIC DIAGRAMS

## 3-1   INTRODUCTION

Schematic diagrams are drawings that show graphically what components are to be used and how these components are to be connected to form a desired circuit. There are two basic types of schematic diagrams: those which use electronic symbols and those which use pictorial symbols. In this chapter we present both types of diagrams.

## 3-2   HOW TO DRAW
SCHEMATIC DIAGRAMS

When drawing schematic diagrams, it is important that the drawing be easy to follow. The symbols should be drawn clearly and neatly with heavy, black lines. All relative data, such as component values and identification numbers, should be located as close as possible to the appropriate symbols [Figure 3-1(a)].

To lay out the diagram clearly, the individual symbols should not be cramped together. A crowded group of symbols is difficult to read and is therefore more likely to cause error. Cramping can be avoided by planning the placement of the symbols ahead of time.

One method that may be used to help assure neat, uncluttered schematic diagrams is to divide the original design sketches into quarters, as shown in Figure 3-2. These quarters need not be equal in area, but may be varied so as to include approximately one-fourth of the symbols. This will give you an idea as to what quarter of the final drawing the symbols should be included in.

After quartering the design sketches, lay out *equal* quarters on the

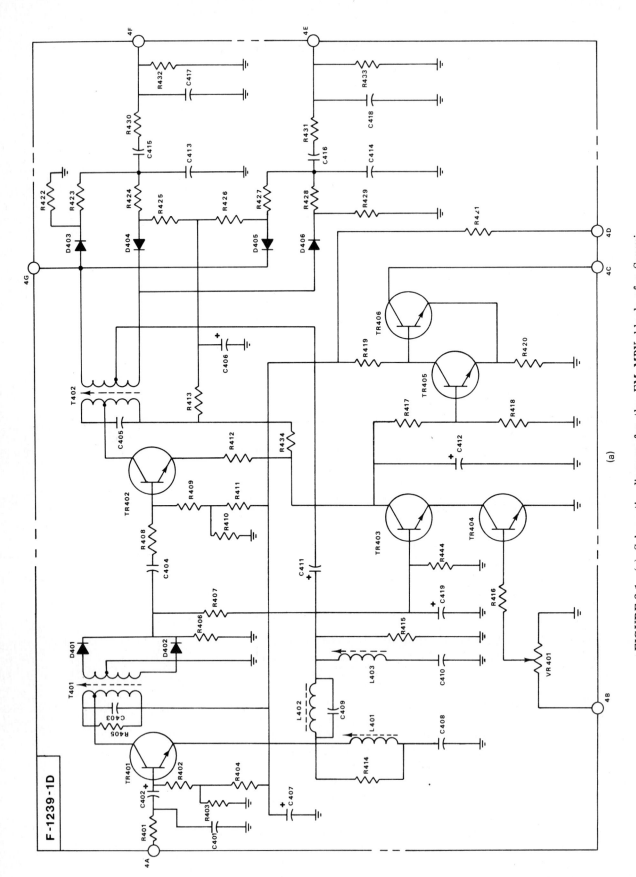

F-1239-1D

FIGURE 3-1 (a) Schematic diagram for the FM MPX block of a Sansui 350A solid-state AM/FM stereo tuner amplifier; (b) schematic diagram parts list for the Sansui 350A shown in part (a). (*Courtesy of Sansui Electric Co., Japan.*)

(a)

**PARTS  LIST– F-1239-1D**

| Part No | Value / Name | | Part No | Value / Name | | |
|---|---|---|---|---|---|---|
| R401 | 1 k | | C412 | 1 UF | 50V | EL |
| R402 | 100 k | | C413 | 680 PF | | |
| R403 | 15 k | | C414 | 680 PF | ±5% 50V | ST |
| R404 | 22 k | | C415 | 0.15 UF | | |
| R405 | 68 k | | C416 | 0.15 UF | ±10% 50V | MY |
| R406 | 100 k | | C417 | 0.006 UF | | |
| R407 | 100 k | | C418 | 0.006 UF | ±5% 50V | MY |
| R408 | 4.7 k | | C419 | 1 UF | 50 | EL |
| R409 | 100 k | | | | | |
| R410 | 2.2 k | | | | | |
| R411 | 22 k | | | | | |
| R412 | 33 k | | TR401 | 2SC711 | | |
| R413 | 220 k | | TR402 | 2SC711 | | |
| R414 | 47 k | | TR403 | 2SC711 | | |
| R415 | 2.2 k  ALL VALUES ARE | | TR404 | 2SC711 | | |
| R416 | 47 k    IN  OHMS | | TR405 | 2SC733 | | |
| R417 | 22 k  ±10% $\frac{1}{4}$ W  CB | | TR406 | 2SC735 | | |
| R418 | 22 k | | | | | |
| R419 | 3.3 k | | | | | |
| R420 | 4.7 | | | | | |
| R421 | 47 | | D401 | IN34A | | |
| R422 | 220 k | | D402 | IN34A | | |
| R423 | 10 k | | D403 | IN34A | | |
| R424 | 10 k | | D404 | IN34A | | |
| R425 | 220 k | | D405 | IN34A | | |
| R426 | 220 k | | D406 | IN34A | | |
| R427 | 10 k | | | | | |
| R428 | 10 k | | | | | |
| R429 | 220 k | | | | | |
| R430 | 56 k | | T401 | 19 kHz Coil | | |
| R431 | 56 k | | T402 | 38 kHz Coil | | |
| R432 | 15 k | | | | | |
| R433 | 15 k | | | | | |
| R434 | 47 k | | | | | |
| | | | L401 | 19 kHz Coil | | |
| | | | L402 | Micro Inductor | | |
| | | | L403 | 67 kHz Coil | | |
| C401 | 68 PF | ±10% 50V | CE | | | |
| C402 | 10 UF | 10V | EL | | | |
| C403 | 10000 PF | ±5% 50V | ST | | | |
| C404 | 0.022 UF | ±10% 50V | MY | | | |
| C405 | 4700 PF | ±5% 50V | ST | | | |
| C406 | 1 UF | 50 V | EL | | | |
| C407 | 47 UF | 25V | EL | | | |
| C408 | 10000 PF | | | | | |
| C409 | 2200 PF | ±5% 50V | ST | | | |
| C410 | 270 PF | | | | | |
| C411 | 10 UF | 25V | EL | | | |

(b)

**FIGURE 3-1**   (cont.)

drawing paper, then draw the components that will take up the most space on the diagram (rotary switches and transistors, for example). By first assuring that these larger symbols are spaced far enough apart so that the diagram does not appear cluttered, the smaller symbols (such as resistors and capacitors) can be easily added.

Component values and/or stock numbers may be included by lettering the data next to the component as shown in Figure 3-3 or by setting up a parts list as shown in Figure 3-1(b). Parts lists have the advantage of allowing purchasers to work from a specific list of components rather than creating their own lists from the schematic. Including

GIVEN DESIGN SKETCH

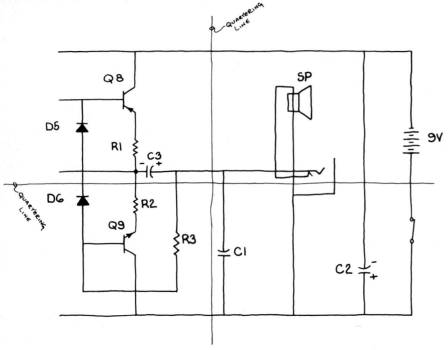

TRANSISTOR RADIO

STEP 1

STEP 2

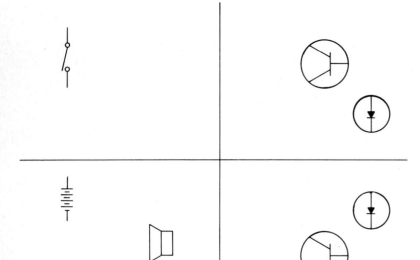

**FIGURE 3-2** How to prepare a schematic diagram.

STEP 3

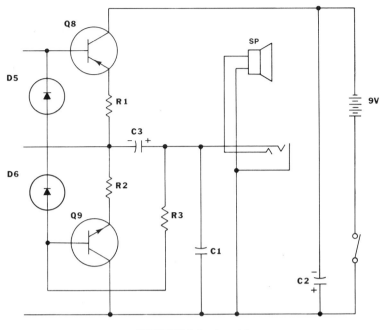

**FIGURE 3-2**   (cont.)

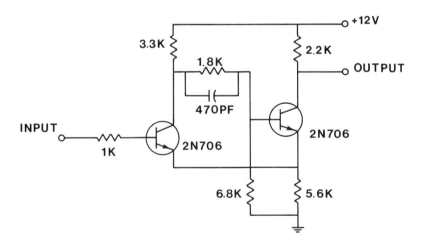

**SCHMITT   TRIGGER**

**FIGURE 3-3**   Schematic diagram.

the data next to the symbols allows technicians to read the diagram more quickly, without having to refer constantly to the table.

When a parts list is used with a schematic diagram, coded part numbers are assigned to each component. For example, all resistors are assigned the letter R followed by a number. The numbers start with 1 and continue as necessary. Figure 3-4 shows a listing of some of the most widely used component letter codes. Figure 3-5 shows an example of a schematic diagram together with an appropriate parts list.

Note that both diodes are labeled DI. This means that both diodes are exactly the same component. All equivalent components are

| COMPONENT | CODE |
|---|---|
| Capacitor | C |
| Diode | D |
| Induction Coil | L |
| Resistor | R |
| Switch | S |
| Transformer | T |
| Transistor | TR or Q |
| Integrated Circuit | Z |

**FIGURE 3-4**  Component letter codes.

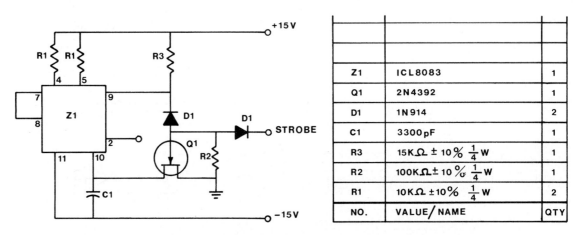

| NO. | VALUE / NAME | QTY |
|---|---|---|
| Z1 | ICL8083 | 1 |
| Q1 | 2N4392 | 1 |
| D1 | 1N914 | 2 |
| C1 | 3300 pF | 1 |
| R3 | $15K\Omega \pm 10\% \frac{1}{4}W$ | 1 |
| R2 | $100K\Omega \pm 10\% \frac{1}{4}W$ | 1 |
| R1 | $10K\Omega \pm 10\% \frac{1}{4}W$ | 2 |
| NO. | VALUE / NAME | QTY |

**FIGURE 3-5**  Schematic diagram with separate parts list.

labeled identically. If, for example, a circuit contained seven 15-kilohm resistors, all would be labeled R with the same number.

Many smaller schematic diagrams do not require parts lists. The value data and manufacturers' numbers are either written directly on the schematic next to the appropriate component or, as shown in Figure 3-6, are coded by include the code interpretations on the body of the drawing. In Figure 3-6, the code interpretations are included in the upper right-hand corner of the schematic diagram.

Schematic diagrams should be drawn in a neat, organized manner. If possible, connector lines should be evenly spaced and major components should be located so that connector lines can be drawn directly (straight) between terminals. This will not only give the drawing a pleasant appearance, but will make it easy to read.

Figure 3-7 shows an example of a poorly laid out schematic diagram. The spacing is uneven and the connector lines ramble. Figure 3-7 also shows a corrected version in which the lines are evenly spaced and direct.

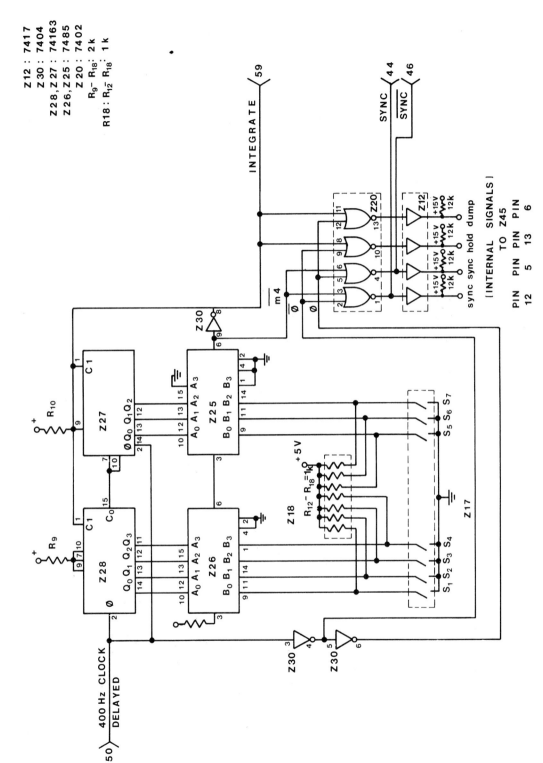

FIGURE 3-6 Schematic diagram parts list included.

65

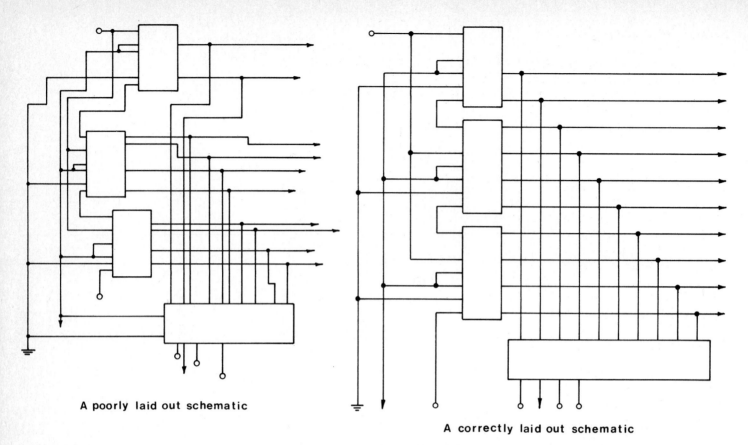

A poorly laid out schematic

A correctly laid out schematic

FIGURE 3-7  How to lay out a schematic diagram.

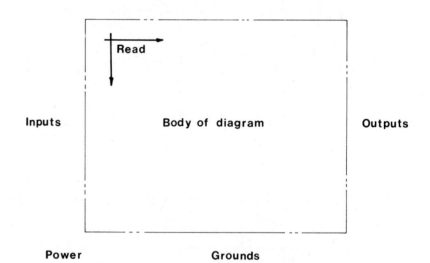

FIGURE 3-8  How to read a schematic diagram.

Schematic diagrams should be drawn following the following general guidelines:

Left side—inputs

Right side—outputs

Lower area—grounds

Lower left—power

66

In general, schematic diagrams are read from left to right and from top to bottom, as shown in Figure 3-8.

*If possible, arrange the schematic diagram so that it can be read from left to right in terms of function. Place together on the drawing symbols that work together to perform a function.*

## 3-3   CROSSOVERS AND INTERSECTIONS

There are two different conventions used to represent crossovers and intersections. The older convention, which is now rarely used, pictures crossovers as loops (semicircles) and intersections as two lines crossing. Figure 3-9 illustrates both types. The newer convention is the dot convention, which indicates intersections with dots and crossovers by crossed lines (see Figure 3-9). With the dot system, when a line joins a line at 90° to form an obvious intersection, the dot may be omitted.

## 3-4   NOTATIONS AND VALUES

Each numerical value given on a schematic diagram must include the units of measure. For example, capacitors are measured in farads, so

**FIGURE 3-9**   How to draw crossovers and intersections on a schematic diagram.

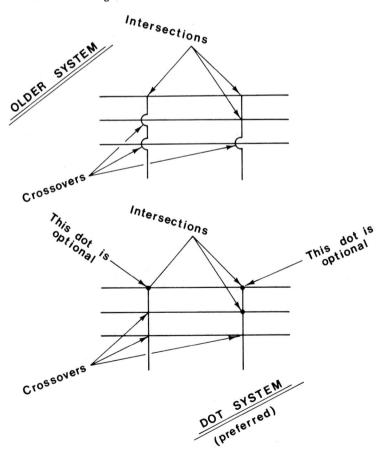

| COMPONENT | UNITS | SYMBOL |
|-----------|-------|--------|
| Resistor | Ohms | $\Omega$ |
| Capacitor | Farads | F |
| Inductor | Henrys | H |
| | Watts | W |
| | Volts | V |
| | Alternating Current | AC |
| | Direct Current | DC |

**FIGURE 3-10**   Notations and values.

**FIGURE 3-11**   Prefixes for multipliers and submultipliers.

## PREFIXES

| Prefix | Symbol | Multiples and Submultiples |
|--------|--------|----------------------------|
| Tera | T | $10^{12}$ |
| Giga | G | $10^{9}$ |
| Mega | M | $10^{6}$ |
| Kilo | K | $10^{3}$ |
| Hecto | H | $10^{2}$ |
| Deka | DA | $10$ |
| Deci | D | $10^{-1}$ |
| Centi | C | $10^{-2}$ |
| Milli | m | $10^{-3}$ |
| Micro | U | $10^{-6}$ |
| Nano | N | $10^{-9}$ |
| Pico | P | $10^{-12}$ |
| Femto | F | $10^{-15}$ |
| Atto | A | $10^{-18}$ |

Examples

2  Picofarads = 2PF = .000 000 000 002 F

4  Milliohms = 4m$\Omega$ = .004 $\Omega$

6  Kilowatts = 6KW = 6000W

$$23.6 \times 10^4 = 236,000$$

$$23.6 \times 10^3 = 23,600$$

$$23.6 \times 10^2 = 2,360$$

$$23.6 \times 10^1 = 236$$

$$23.6 = 23.6$$

$$23.6 \times 10^{1} = 2.36$$

$$23.6 \times 10^{2} = .236$$

$$23.6 \times 10^{3} = .0236$$

$$23.6 \times 10^{4} = .00236$$

**FIGURE 3-12**   How to read multiples of 10.

each capacitor value must be accompanied by the farad unit of measure: for example, 2 farads, 0.4 farad.

Each unit of measure has a symbol to make it faster to print on a drawing. Figure 3-10 lists the units of measure and their symbols for components called for most often on schematic diagrams.

Units of measure often include a multiplier. A kilowatt, for example, means 1000 watts, or $1 \times 10^3$. "Kilo" is a multiplier which when used as a prefix to a unit of measure means that the unit of measure is to be multiplied by 1000. A megaton means 1,000,000 tons. "Mega" means 1,000,000, or $1 \times 10^6$. Figure 3-11 shows additional examples.

A submultiplier is a multiplier that has a negative exponent. A microfarad is equal to 0.001 farad. "Micro" means to multiply by 1/1000, or $1 \times 10^{-3}$. "Pico" means 1/1,000,000,000,000, or 0.000 000 000 001, or $1 \times 10^{-12}$. We can see how much easier the symbols for prefixes make writing values and units of measure. A picofarad would be written on a drawing as: 1 PF.

It is sometimes difficult to keep track of all the decimal places involved in using multipliers. How many decimal places (zeros) are added to 23.6 when the number is written $23.6 \times 10^5$? For positive exponents (the small numbers written above and to the right of the 10's), add decimal places to the right of the decimal point equal to the exponential value; therefore, $23.6 \times 10^5$ equals 2,360,000. For negative exponents, add decimal places to the left; therefore, $23.6 \times 10^{-5}$ equals 0.000 236. Figure 3-12 illustrates this method.

## 3-5   BASIC CIRCUITS

Figure 3-13 illustrates some of the more common circuit configurations. These basic configurations are found as part of many more sophisticated circuit designs. The drafter should be familiar enough with these basic circuits to recognize them when they appear in design sketches.

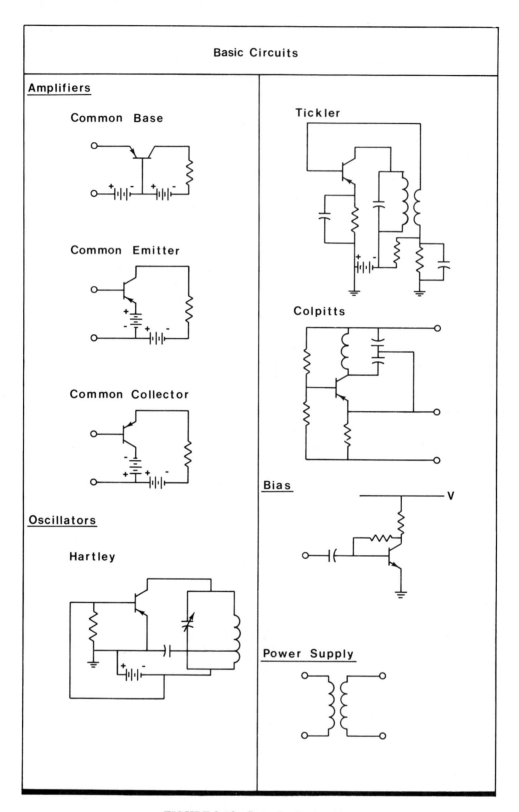

**FIGURE 3-13** Some basic circuits.

## 3-6    INTEGRATED CIRCUITS IN SCHEMATIC DIAGRAMS

Integrated circuits (ICs) are used extensively in many different types of schematic diagrams. They usually appear as rectangles, squares, or equilateral triangles. See Section 2-4 for an explanation of the graphic symbols used to represent ICs.

Pin numbers on an IC are always assigned consecutively starting with 1. Figure 3-14 shows a dual-type D flip-flop IC device with the pins arranged in order. Figure 3-14 shows the same IC as it might appear on a schematic diagram. Note that the pins are not arranged in numerical order but are arranged by function. Input pins 5 and 9 are on the left, output pins 1 and 13 on the right, and other functions in the middle. Grounds 4, 6, 8, and 10 are located toward the bottom of the diagram. Pins not used are omitted.

Designers and engineers usually arrange IC pins to follow the design function. This makes it easier to read and understand the diagram.

IC devices may also be arranged on a schematic diagram to show the internal function as related to the rest of the circuit. Figure 3-15 shows a wide-bandwidth dual-JFET-input operation amplifier: first as it would appear with the pins in consecutive order, then as it might appear as part of a circuit on a schematic diagram. The pins are still numbered, although two amplifier symbols are used rather than the rectangular IC symbol. The notation ½ 353 states the IC device's manufacturer's part number, 353, with the ½ implying that the amplifier shown is only one-half of the IC device.

When preparing schematic diagrams, always arrange the pins of an IC device to follow the design function. The shorter edges may also be

**FIGURE 3-14**  Not all pins need be used or drawn in consecutive order.

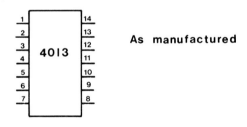

As manufactured

As could appear on

a schematic diagram

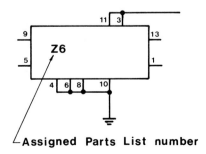

Assigned Parts List number

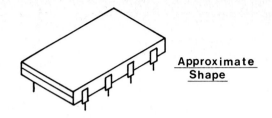

Approximate
Shape

Pin Connection

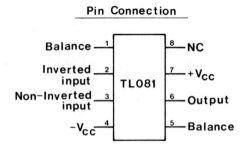

As represented on a schematic diagram

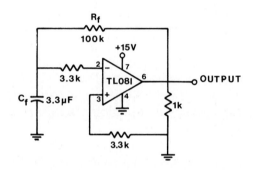

FIGURE 3-15  Sample schematic diagram that uses an IC.

used for pin locations even though no pins are actually located on the edges.

IC amplifier devices are manufactured in different shapes, but are always drawn on schematic diagrams as equilateral triangles. Figure 3-16 shows a JFET-input operation amplifier. The actual amp chip is rectangular, with four pins on each of the two longer sides. A manufacturer's description of the chip would also be rectangular as shown. However, when used as a triangle, as shown in Figure 3-16, pins are represented by straight lines located around the triangle.

## 3-7   PICTORIAL SCHEMATIC DIAGRAMS

Pictorial schematic diagrams are schematic diagrams that use the pictorial symbols defined in Chapter 2 instead of the ANSI symbols. These schematics are useful to persons not familiar with the standard symbols, but they lack the explicit detail required by designers. Figure 3-17 shows a schematic diagram and the same diagram in pictorial form. Pictorial schematics are also helpful in the preparation of printed circuit drawings, which are covered in Chapter 6.

72

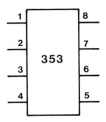

**As could appear on a schematic diagram**

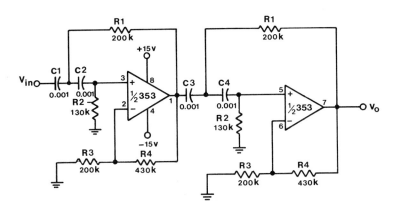

**FIGURE 3-16** Split IC as drawn in a schematic diagram.

**FIGURE 3-17** Schematic diagram and the same circuit as a pictorial schematic diagram.

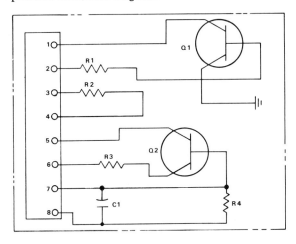

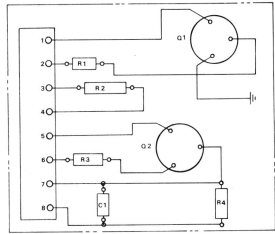

## PROBLEMS

**3-1**  Express the following values in words and then as they would be printed on a drawing.
  1.  0.003 amp
  2.  0.000 000 004 farad
  3.  1000 ohms
  4.  10,000 watts
  5.  2,500,000 ohms
  6.  0.000 000 000 025 farad
  7.  4,000,000 ohms
  8.  0.000 003 farad
  9.  0.025 amp
  10.  1500 volts

**3-2**  Write the numerical equivalents of the following. Be sure to include the units of measure.
  1.  4 PF
  2.  3 KV
  3.  1.5 mA
  4.  2 KΩ
  5.  6 KW
  6.  1.5 Ω
  7.  3 NF
  8.  1.75 MΩ
  9.  2 UA
  10.  1 MW

**3-3**  Redraw the Colpitts oscillator circuit shown in Figure P3-3 and replace the letters R1, R2, R3, C1, C2, L1, and Q1 with the appropriate symbol and value. The letters are defined as follows:

| Letter | Component  | Value   |
|--------|------------|---------|
| R1     | Resistor   | 12 KΩ   |
| R2     | Resistor   | 8.2 KΩ  |
| R3     | Resistor   | 1.5 KΩ  |
| C1     | Capacitor  | 0.10 PF |
| C2     | Capacitor  | 0.047 PF|
| L1     | Inductor   | 10 MH   |
| Q1     | Transistor | 2N2926  |

**FIGURE P3-3**

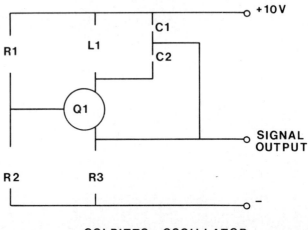

**COLPITTS OSCILLATOR**

3-4    Redraw the basic logic circuit shown in Figure P3-4 and add the following values.

R1    10 KΩ
R2    4.7 KΩ
Q1    2N708    (NPN transistor)

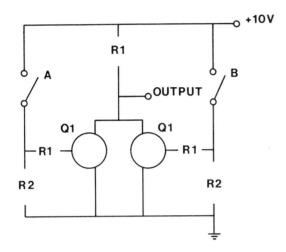

**BASIC  LOGIC  CIRCUIT**

Circuit courtesy of General Electric Co.

**FIGURE P3-4**

3-5    Redraw the power supply circuit shown in Figure P3-5.

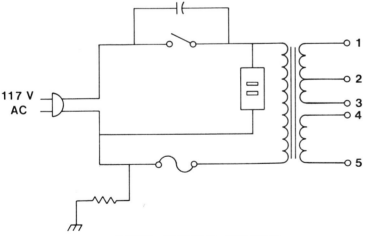

**POWER  SUPPLY  CIRCUIT**

Circuit courtesy of General Electric Co.

**FIGURE P3-5**

3-6    Redraw the AM broadcast band transmitter shown in Figure P3-6 and substitute as follows.
  1.  670 KC    CRYSTAL
  2.  150 KΩ
  3.  2N170    (NPN transistor)

4. 300 PF
5. Ground
6. Ground
7. 2N170     (NPN transistor)
8. 0.002 PF
9. Ground
10. Inductor     (10 MH RFC)
11. 1 KΩ
12. Battery 6 V
13. Ground
14. 0.001 PF
15. 0.003 PF
16. L3
17. Ground
18. 10 mF, 10 V
19. 0.001 PF
20. Antenna
21. Chassis ground

Add both symbols and values.

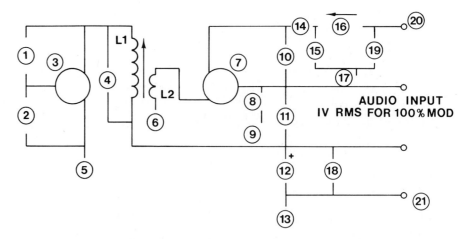

**AM  BROADCAST  BAND  TRANSMITTER**

**FIGURE P3-6**

3-7   Redraw the schematic diagram for the 6-volt phono amplifier
shown in Figure P3-7 and substitute as follows.
1. Letter in the words CRYSTAL CARTRIDGE
2. 10 KΩ
3. 8 mF
4. 2N323     (transistor)
5. 150 KΩ
6. 0.5 mF
7. 6.8 KΩ
8. R1     (tapped resistor)
9. 1 KΩ
10. 0.2 mF
11. 0.02 mF
12. 10 KΩ
13. 0.05 mF

14.  R3      (tapped resistor)
15.  R2      (tapped resistor)
16.  100 KΩ
17.  0.1 mF
18.  220 KΩ
19.  2N323     (transistor)
20.  6 V, 10 mF
21.  2.2 KΩ
22.  10 mF
23.  4.7 KΩ
24.  2N323     (transistor)
25.  47 KΩ
26.  1.5 KΩ
27.  330 Ω
28.  6 V, 50 mF
29.  33 KΩ
30.  T1      (iron-core transformer)
31.  220 Ω
32.  6 V, 50 mF
33.  6 V, 50 mF
34.  Ground symbol
35.  1.2 KΩ
36.  2N1415     (transistor)
37.  33 Ω
38.  2N1415     (transistor)

**FIGURE P3-7**

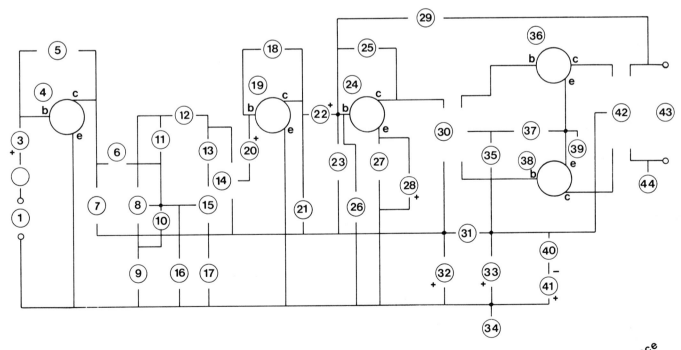

SIX  VOLT  PHONE  AMPLIFIER

Add notes here

Add performance data here

39. Ground symbol
40. Single-throw switch
41. 6-V battery
42. T2     (iron-core transformer)
43. Letter in the words TO SPEAKER
44. Ground symbol

All transistors are PNP types. In addition, add the following notes and performance data.

*NOTES:*
R1—Bass Control—50 K Linear Taper
R2—Treble Control—50 K Linear Taper
R3—Volume Control—10 K Audio Taper
T1—Driver Transformer—PR1 2 K/sec 1.5 K, C.T.
T2—Output Transformer—PR1 100 $\Omega$/sec V.C. (3.2, 8, 16 $\Omega$)
All resistors ½ watt

*PERFORMANCE DATA*
Maximum power out at 10% distortion          300 mW
Distortion at 100 mW                                    60 Hz—3%
                                                                    1.0 RHz—1.5%
                                                                    5.0 RHz—3.0%

3-8  Much of the work assigned beginning drafters is in the form of freehand sketches. An engineer will present freehand

**FIGURE P3-8**

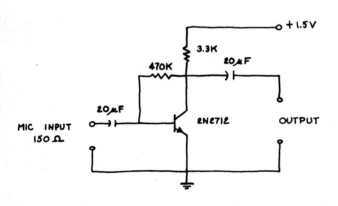

MIC PREAMPLIFIER

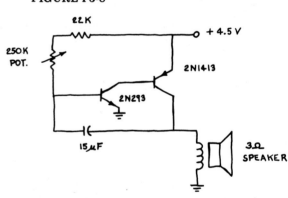

METRONOME

EMITTER TRIGGERING

design sketches and have a drafter prepare the finished schematic drawings from these sketches. Study the design sketches in Figure P3-8 and redraw them using drawing instruments.

3-9   Redraw the Colpitts circuit of Figure P3-3 as a pictorial schematic.

3-10  Redraw the MIC preamplifier circuit of Figure P3-8(a) as a pictorial schematic.

3-11  Redraw the pictorial schematics in Figures P3-11 through through   P3-15 as schematic diagrams using symbols. 3-15

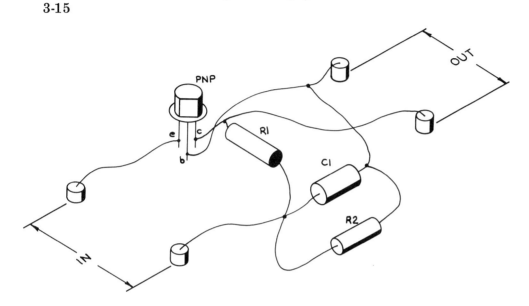

**FIGURE P3-11**

**FIGURE P3-12**

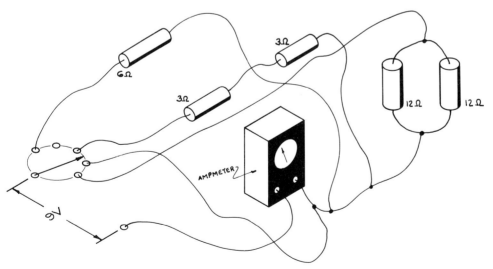

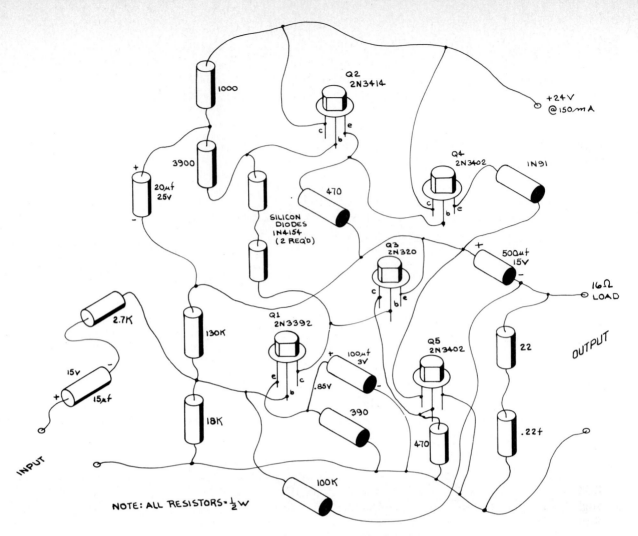

**FIGURE P3-13**

**FIGURE P3-14**

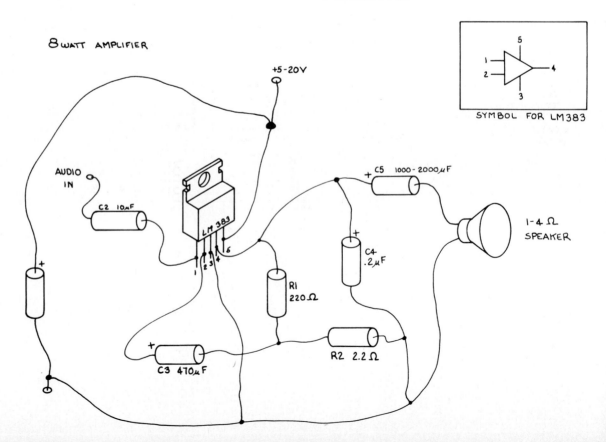

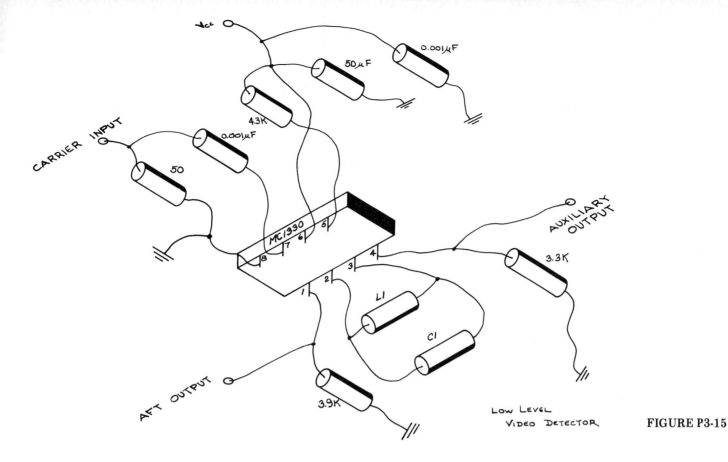

Low Level
Video Detector

**FIGURE P3-15**

3-16 through 3-20    Figures P3-16 through P3-20 are designers' sketches of various circuits. In each problem, the sketch is to be redrawn using instruments or, if available, a computer graphics system. Problems 3-19 and 3-20 should be drawn on 11 × 17 inch paper.

**FIGURE P3-16**

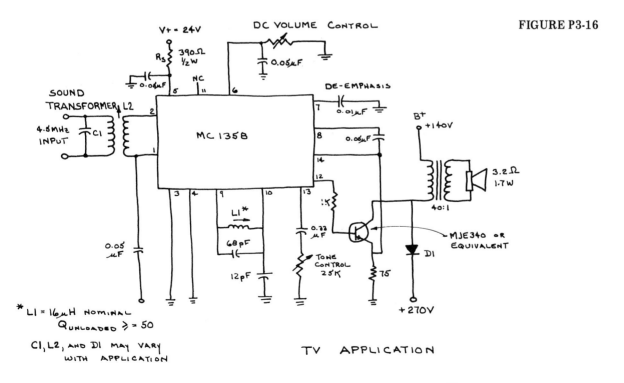

TV APPLICATION

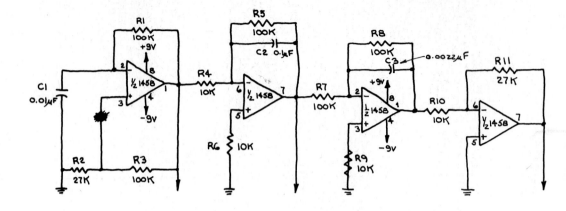

FUNCTION   GENERATOR

**FIGURE P3-17**

**FIGURE P3-18**

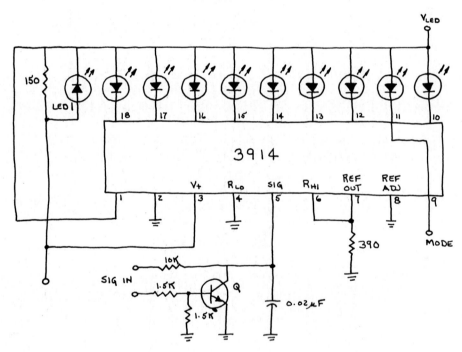

EXCLAMATION POINT
DISPLAY

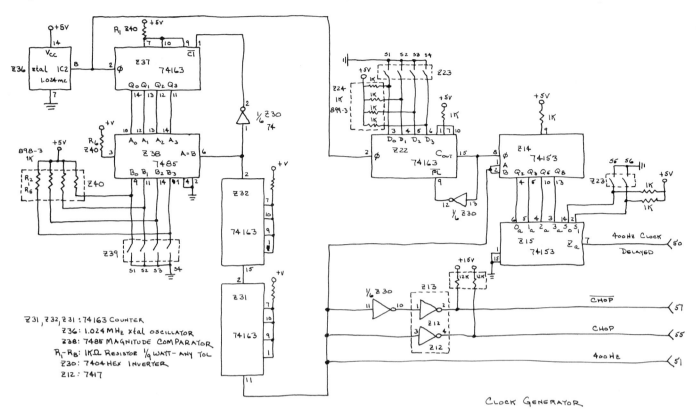

**FIGURE P3-19**

**FIGURE P3-20**

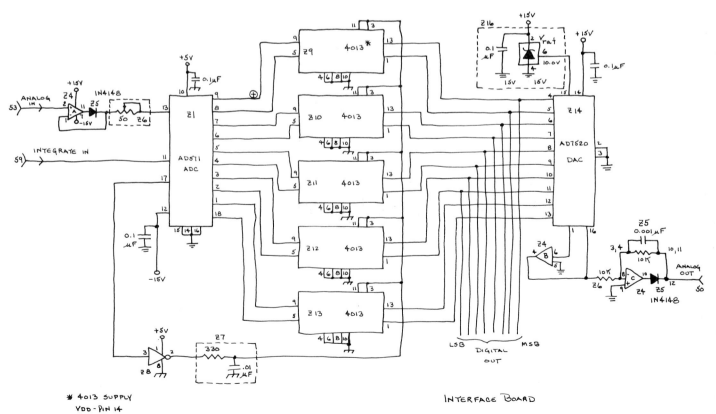

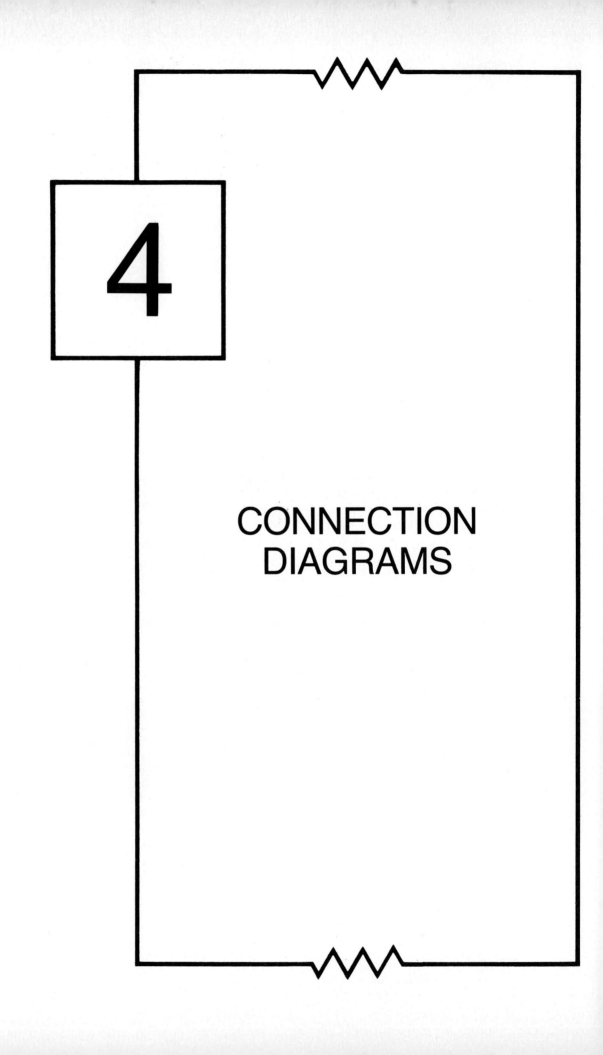

# 4

## CONNECTION DIAGRAMS

## 4-1 INTRODUCTION

Four different types of connection diagrams are described in this chapter. *Connection diagrams* are drawings that define how various components of a system are to be wired together. They are used most often in conjunction with assembly or maintenance instructions and as design layouts.

Each of the four types of connection diagrams discussed has advantages and disadvantages in its preparation and use. As you study how to draw the diagrams, try to become aware of these advantages and disadvantages so that you can learn which type is best suited for the requirements of each drawing.

## 4-2 POINT-TO-POINT DIAGRAMS

*Point-to-point diagrams* show the terminal connection location and routing path of every wire used in a system. Wires are not shown in bundles, nor do they use destination codes or tables. Each line drawn represents one wire which is graphically shown starting at one point and ending at another.

Point-to-point diagrams are very useful in design work because they enable the reader to follow directly the path of each wire. However, for large drawings, which contain many wires, point-to-point diagrams can become confusing and difficult to read accurately.

To draw a point-to-point diagram (see Figure 4-1):

STEP 1 - Draw Components                              STEP 2 - Add Conductor Paths

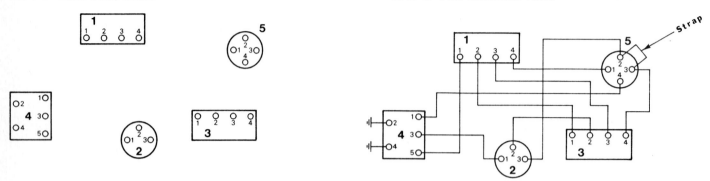

STEP 3 - Add Wire Colors

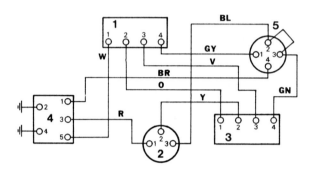

FIGURE 4-1   How to prepare a point-to-point diagram.

1. Draw each component, locating it in the same position as that in which it is located in the installation. Label all components and terminals.

2. Draw in the wire paths. Wires that connect terminals on the same component are called *straps*.

3. If required, assign colors to each wire. A table of wire color codes is included in Appendix E.

Some drafters prefer to rearrange the terminals numbers so that they can make neater line patterns which are easier to follow. Figure 4-2 illustrates this technique. Note how, by changing the sequence of the terminal numbers, the diagram becomes much neater in appearance and much easier to follow.

FIGURE 4-2   Example of how rearranging the terminal numbers can be used to clarify a point-to-point diagram.

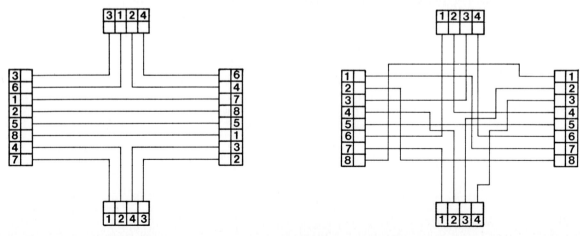

## 4-3   BASELINE DIAGRAMS

*Baseline diagrams* are drawings that feed all wires into one central line called a baseline. They present the wiring in a well-organized, easy-to-follow format, but have the disadvantage that they do not show components in their correct physical positions. Baseline drawings are particularly useful in presenting large, many-component diagrams in a size suitable for book use. They are used most often for maintenance and assembly manuals.

To draw a baseline diagram (Figure 4-3 illustrates):

1.  Draw a light horizontal line across the center of the paper. This line is called the baseline.
2.  Draw half of the components above the baseline and half below. Label all components and terminals.
3.  Connect all used terminals to the baseline by drawing lines directly from the terminals to the baseline. If a direct path is not possible, make all directional changes at $90°$. The baseline may not be by-passed. All wires must be connected to the baseline. (Straps need not be connected to the baseline.)
4.  Label all wires using a destination code.

**FIGURE 4-3**   How to prepare a baseline diagram.

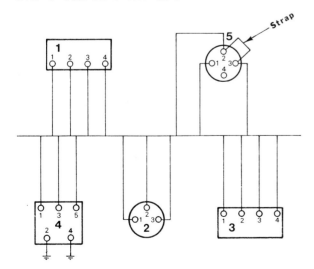

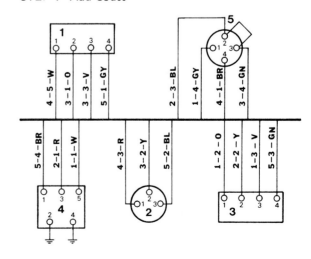

**Destination Code**

**Component No — Terminal No — Wire Color**

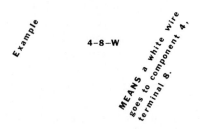

**FIGURE 4-4**   Destination code.

The destination code consists of letters and numbers that identify the component number and terminal number to which the wire is to be attached, and the color of the wire. See Figure 4-4 for a further explanation of the destination code.

5.   Darken in the baseline, making it a very heavy, black line.

## 4-4   HIGHWAY DIAGRAMS

*Highway diagrams* combine groups of wires running along similar paths into bundles called *highways*. The components are located in the same relative positions as they would be located in the actual component setup. Of the different types of component diagrams shown in this chapter, highway diagrams most nearly duplicate the wiring configurations as they appear in the final installation.

Figure 4-5 is a point-to-point diagram that we wish to redraw as a highway diagram. To draw the highway diagram (Figure 4-6 illustrates):

**FIGURE 4-5**   Point-to-point diagram that is to be redrawn as a highway diagram and as a lineless diagram.

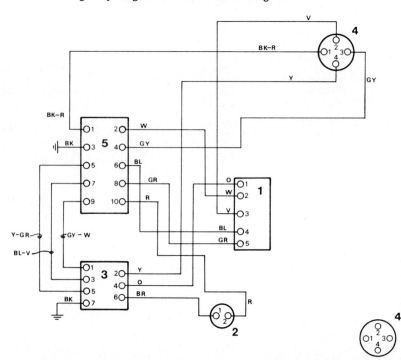

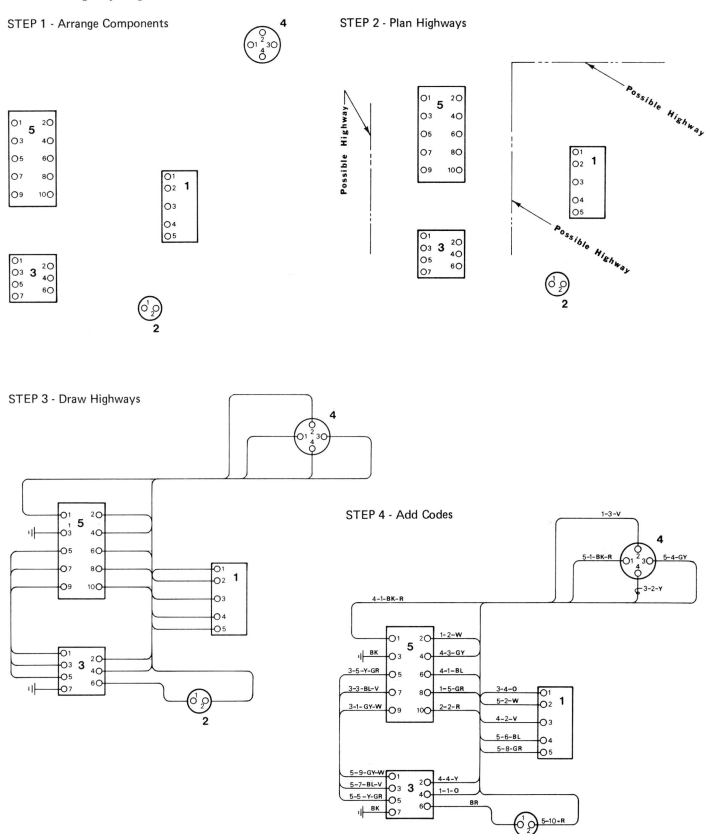

**FIGURE 4-6**  How to prepare a highway diagram.

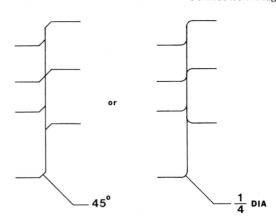

FIGURE 4-7  Wire turns and intersections on a highway drawing may be drawn as either an arc or as a line slanted at 45°.

1.  Draw the components, locating them in the same positions as those they occupy in the final installation. Identify all components and terminals.

2.  Study the drawing and with light construction lines, identify all possible areas where wires may be bundled. These bundles are called highways.

3.  Draw very light lines from the appropriate terminals to the highways. Where an individual wire joins the highway, use either an arc or a line slanted at 45°, as shown in Figure 4-7. The direction of the arc or slanted line should be in the direction the wire is headed. For example, the arc showing the intersection of the wire from terminal 10 of component 5 to the highway turns down as the wire is headed for component 2, whereas the wire from terminal 3 of component 1 turns up as the wire is headed for component 4. *Note that unlike baseline diagrams, highways may be bypassed. The wire from terminal 6 of component 3, for example, may be drawn directly to terminal 1 of component 2. It need not be included as part of a highway.*

4.  Darken all lines and label each wire with a destination code: component number–terminal number–wire color. Ground wires need only be labeled with their colors and the symbol for ground.

## 4-5   LINELESS DRAWINGS

*Lineless drawings* are component drawings that do not show any wiring paths but instead define the wiring paths by using a wiring table. Lineless diagrams are particularly useful for clarifying large complex drawings where the number of wires is so great that several sheets of drawings would be required to include the entire pattern. The disadvantage of lineless diagrams is that because they omit all line patterns, they present an incomplete representation of what the final wiring setup will look like.

Figure 4-5 is a point-to-point diagram that we wish to redraw as a lineless diagram. To draw the lineless diagram (Figure 4-8 illustrates):

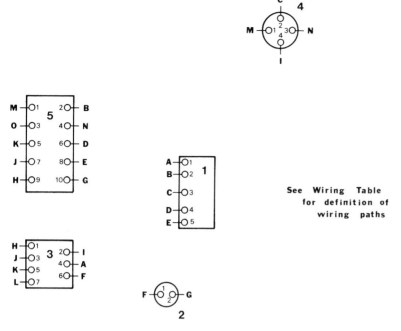

| WIRING TABLE | | | |
|---|---|---|---|
| WIRE I.D. | WIRE COLOR | FROM | TO |
| A | O | 1–1 | 3–4 |
| B | W | 1–2 | 5–2 |
| C | V | 1–3 | 4–2 |
| D | BL | 1–4 | 5–6 |
| E | GR | 1–5 | 5–8 |
| F | BR | 2–1 | 3–6 |
| G | R | 2–2 | 5–10 |
| H | GY–W | 3–1 | 5–9 |
| I | Y | 3–2 | 4–4 |
| J | BL–V | 3–3 | 5–7 |
| K | Y–GR | 3–5 | 5–5 |
| L | BK | 3–7 | GRD |
| M | BK–R | 4–1 | 5–1 |
| N | GY | 4–3 | 5–4 |
| O | BK | 5–3 | GRD |

See Wiring Table for definition of wiring paths

**FIGURE 4-8**   Example of a lineless diagram.

1. Draw the components of the system. If possible, locate the components on the drawing in positions that approximate their actual physical locations, and number the terminals. Each component must be labeled.

2. Prepare a wiring chart such as the one shown in Figure 4-9. The chart should include wire identification, where the wire starts (component and terminal), and where the wire ends (component and terminal). The chart may also include wire size and wire color. In this example, the wires were identified by letters and the components by numbers. Component names could also have been used.

3. Draw a short line from each terminal used and label it with the appropriate wire identification letter.

**FIGURE 4-9**   Example of a wiring table that would accompany a lineless diagram.

| WIRING TABLE | | | |
|---|---|---|---|
| WIRE I.D. | WIRE COLOR | FROM | TO |
| A | G | 1–1 | 3–2 |
| | | | |
| | | | |
| | | | |
| | | | |
| | | | |

*Sample Callout*

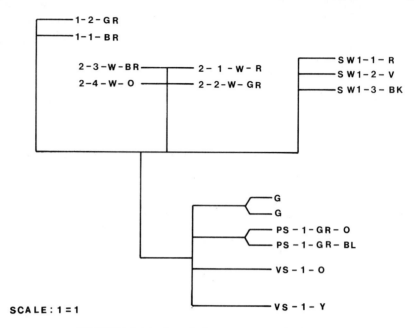

SCALE: 1 = 1

**FIGURE 4-10**  Sample harness assembly diagram.

## 4-6   HARNESS ASSEMBLY DRAWINGS

A *harness assembly drawing* defines the pattern for wiring in an electrical assembly and how the wires are to be bundled together. Figure 4-10 is an example of a harness assembly drawing.

Wires are bound together in groups called harnesses, as shown in Figure 4-11. Wiring harnesses are made outside the assembly. This makes it easier to bundle the wires (the components are not in the way) and also helps to prevent damage to the assembly.

Harness assembly drawings are drawn at a 1:1 scale: in other words, at their actual size. This is done so that the drawing can be used to help construct the harness.

To create a harness assembly drawing (see Figure 4-12):

**FIGURE 4-11**  Harnesses.

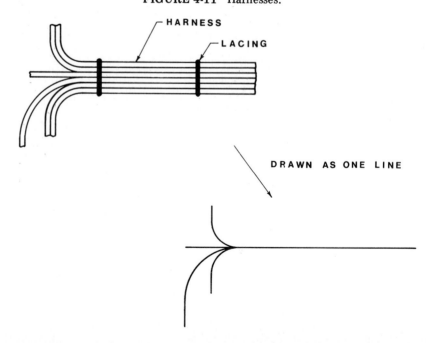

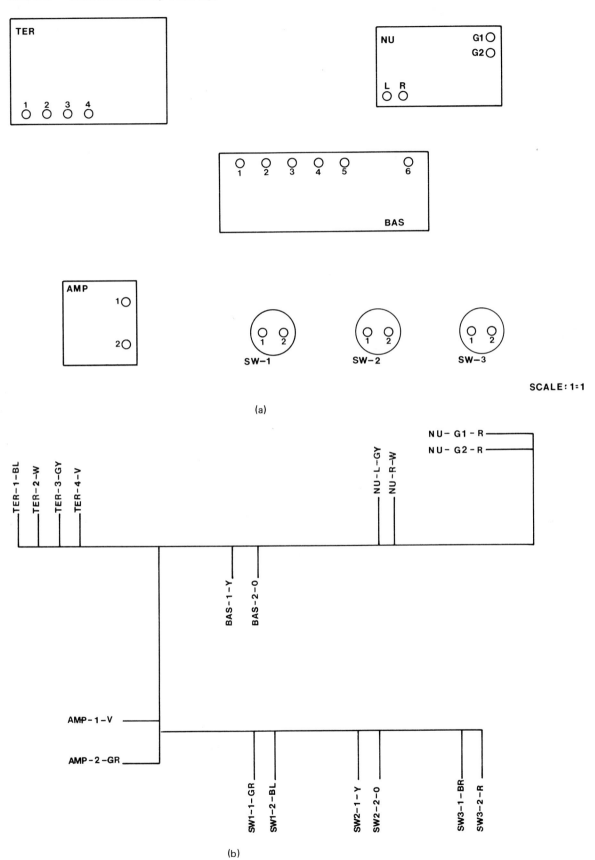

FIGURE 4-12  How to create a harness assembly diagram.

| FROM LOCATION | LEAD TERMINATION (STRIPPED) | COLOR OF INSULATION | LEAD LENGTH (INCHES) | TO LOCATION | LEAD TERMINATION (STRIPPED) |
|---|---|---|---|---|---|
| TER – 1 | $\frac{3}{4}$ | BL | $8\frac{3}{16}$ | SW1 – 2 | $\frac{3}{4}$ |
| TER – 2 | $\frac{3}{4}$ | W | $6\frac{13}{16}$ | NU – R | $\frac{3}{4}$ |
| TER – 3 | $\frac{3}{4}$ | GY | $6\frac{1}{8}$ | NU – L | $\frac{3}{4}$ |
| TER – 4 | $\frac{3}{4}$ | V | $10\frac{9}{16}$ | NU – G2 | $\frac{1}{2}$ |
| AMP – 1 | $\frac{3}{4}$ | O | $5\frac{3}{16}$ | BAS – 2 | $\frac{1}{2}$ |
| AMP – 2 | $\frac{1}{2}$ | GR | $2\frac{5}{16}$ | SW1 – 1 | $\frac{3}{4}$ |
| SW1 – 1 | $\frac{3}{4}$ | GR | $2\frac{5}{16}$ | AMP – 2 | $\frac{1}{2}$ |
| SW1 – 2 | $\frac{3}{4}$ | BL | $8\frac{3}{16}$ | TER – 1 | $\frac{3}{4}$ |
| SW2 – 1 | $\frac{1}{2}$ | Y | $8\frac{5}{8}$ | BAS – 1 | $\frac{1}{2}$ |
| SW2 – 2 | $\frac{1}{2}$ | BR | 2 | SW3 – 1 | $\frac{3}{4}$ |
| SW3 – 1 | $\frac{3}{4}$ | BR | 2 | SW2 – 2 | $\frac{1}{2}$ |
| SW3 – 2 | $\frac{3}{4}$ | R | $18\frac{1}{4}$ | NU – G1 | $\frac{1}{2}$ |
| NU – L | $\frac{3}{4}$ | GY | $6\frac{1}{8}$ | TER – 3 | $\frac{3}{4}$ |
| NU – R | $\frac{3}{4}$ | W | $6\frac{13}{16}$ | TER – 2 | $\frac{3}{4}$ |
| NU – G1 | $\frac{1}{2}$ | R | $18\frac{1}{4}$ | SW3 – 2 | $\frac{3}{4}$ |
| NU – G2 | $\frac{1}{2}$ | V | $10\frac{9}{16}$ | TER – 4 | $\frac{3}{4}$ |
| BAS – 1 | $\frac{1}{2}$ | Y | $8\frac{5}{8}$ | SW2 – 1 | $\frac{1}{2}$ |
| BAS – 2 | $\frac{1}{2}$ | O | $5\frac{3}{16}$ | AMP – 1 | $\frac{3}{4}$ |

(c)

FIGURE 4-12   (cont.)

1. Create a layout drawing, using a 1:1 scale, of all components and connections in the assembly. Draw all components using their actual size and locate them using the actual distance from each other. Identify all components.

2. Locate and identify all (even those not used) terminals on all components [see Figure 4-12(a)].

3. Place a second sheet of drawing paper over the layout and sketch in the wire paths. Any transparent paper is acceptable.

4. Using freehand sketches, combine the wire paths into groups as shown. Several attempts are usually required before a final arrangement is obtained.

5. Remove the freehand sketches and place another sheet of paper (drawing film) over the layout. Use the final freehand sketch as a guide, and using instruments, draw the required harness pattern [see Figure 4-12(b)].

6. Identify all leads using the destination code explanation in Figure 4-4. Define the wire colors.

7. Create a table as shown. All lengths may be measured off the harness drawing layout [see Figure 4-12(c)].

There are many different pieces of information that could be listed in a harness table, including lacing termination distance, American wire gage (AWG) number, insulation type, type of wire, and lead termination requirements. The table presented here gives only the most basic information.

## PROBLEMS

4-1   Given the point-to-point diagram in Figure P4-1, redraw the diagram as:
   a.   A highway diagram
   b.   A baseline diagram
   c.   A lineless diagram
   d.   A harness assembly drawing
   In preparing the harness assembly drawing, assume that all components are drawn and located at a 1:1 scale.

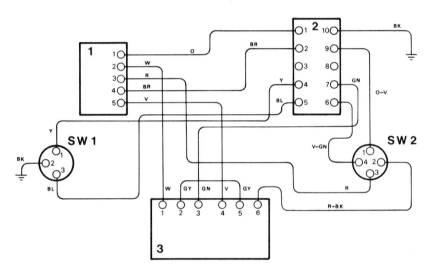

**FIGURE P4-1**

4-2   Figure P4-2 is an illustration of a home stereo system which includes two speakers, an amplifier, and a turntable. Redraw the system as:
   a.   A highway diagram
   b.   A baseline diagram
   c.   A lineless diagram
   d.   A point-to-point diagram
   Assign component numbers, terminal numbers, and wire colors as needed. Wire color abbreviations are given in Appendix E.

4-3   Your company is working on a new electronic system. A prototype has been set up in the lab and you have been asked to study the prototype setup (pictured in Figure P4-3) and prepare a wiring diagram. Redraw the system as:
   a.   A highway diagram

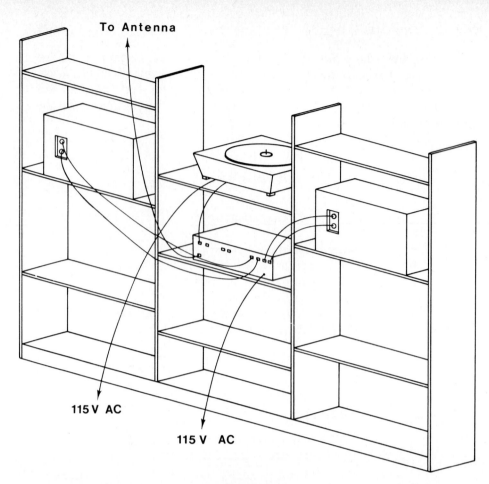

To Antenna

115 V AC

115 V AC

**FIGURE P4-2**

**FIGURE P4-3**

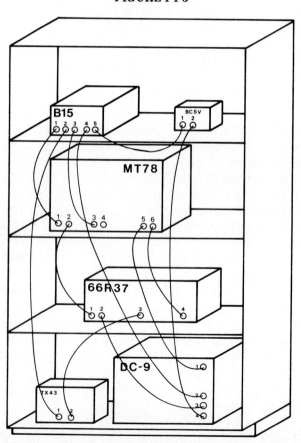

B15

BC 5V

MT78

66R37

DC-9

TX43

96

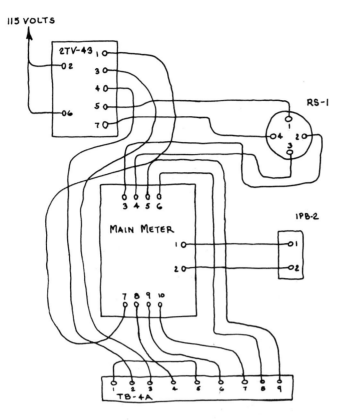

DESIGN SKETCH - DIPLEX SYSTEM

**FIGURE P4-4**

    **b.**  A baseline diagram
    **c.**  A lineless diagram
    **d.**  A point-to-point diagram
    **e.**  A harness assembly drawing

**4-4**    Figure P4-4 is an engineer's design sketch. Redraw the sketch as:
    **a.**  A highway diagram
    **b.**  A baseline diagram
    **c.**  A lineless diagram

**4-5**    Figures P4-5 through P4-8 are representative of the types of
**through**  design sketches for which drafters are often asked to create
**4-8**    appropriate drawings. For each problem, create one of the following:
    **a.**  A highway diagram
    **b.**  A baseline diagram
    **c.**  A lineless diagram
    **d.**  A point-to-point diagram
    **e.**  A harness assembly drawing
    Problems 4-7 and 4-8 are to be drawn on 11 X 17 inch paper. Assign wire colors as necessary. If preparing a harness assembly drawing, assume that all components are drawn and located at a 1:1 scale.

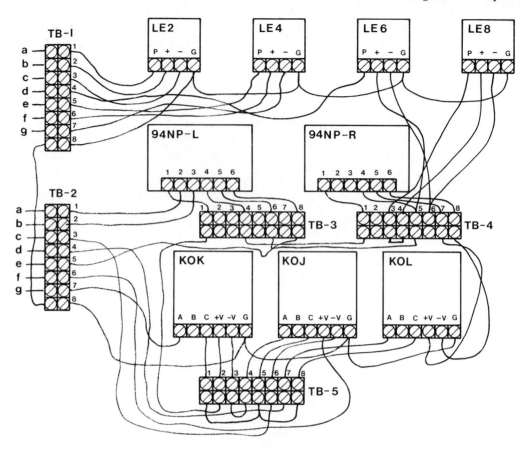

**FIGURE P4-5**

**FIGURE P4-6**

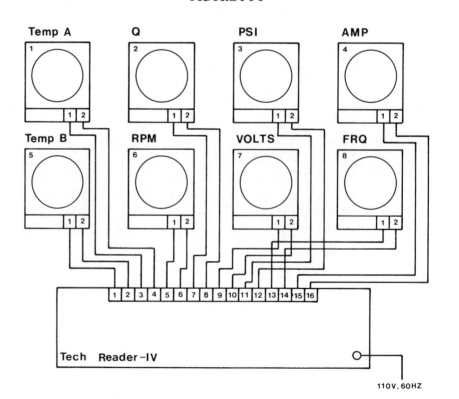

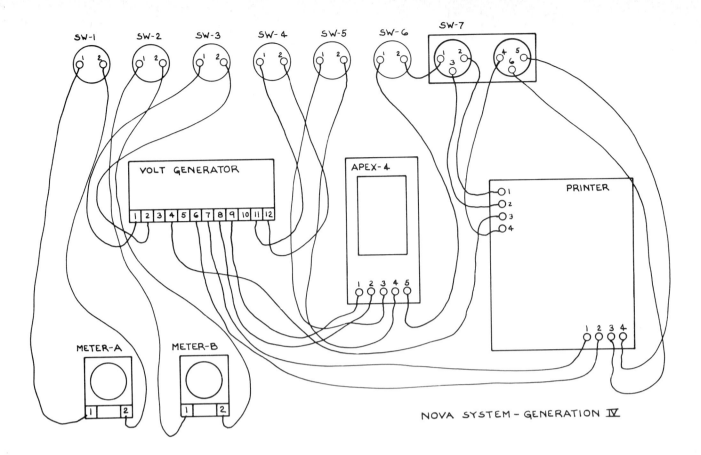

NOVA SYSTEM - GENERATION IV

**FIGURE P4-7**

**FIGURE P4-8**

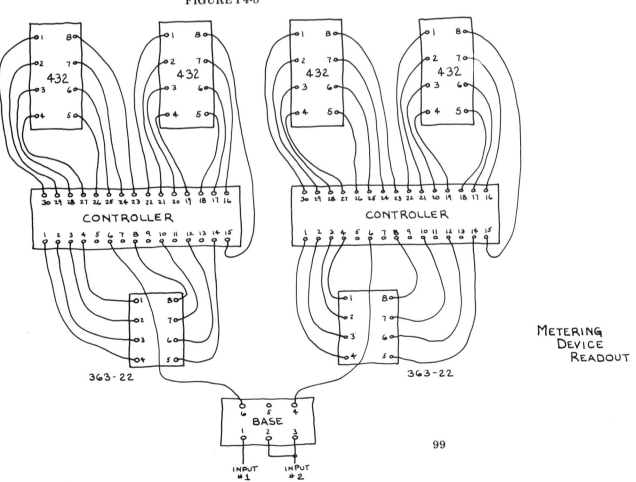

METERING
DEVICE
READOUT

# 5

# BLOCK
# AND LOGIC
# DIAGRAMS

## 5-1 INTRODUCTION

In this chapter we explain how to draw and interpret block and logic diagrams. As in previous chapters, the drawing fundamentals required are explained in detail, but in addition, a discussion of the use of each type of diagram is also included.

## 5-2 BLOCK DIAGRAMS

*Block diagrams* are a way to express graphically the relationships among a series of elements. They are used in the electronic and electrical fields, and in almost every other technical area of study: business, mathematics, engineering, to name just a few. Figure 5-1 is an example of a block diagram that illustrates a stereo system.

To draw a block diagram, no special block size or shape is required, although rectangular blocks are generally used. Usually, all blocks on a diagram are drawn the same size and all lines are drawn the same thickness. If special emphasis is desired, a larger block or a thicker line or a combination of both is drawn. Note how the amp in Figure 5-1 is emphasized by a thicker line.

Block diagrams are set up to be read from left to right and from top to bottom. If more than one line of blocks is required, the second line must be located under the first row and the line of flow must be returned from the end of the first row to the beginning of the second row, as shown in Figure 5-2. If one of the blocks represents an element that is not normally part of the series, it can be drawn using hidden lines, as was done with the earphones in Figure 5-1.

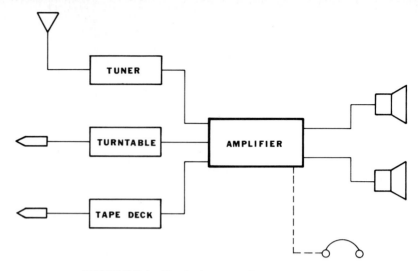

FIGURE 5-1  Block diagram of a stereo system.

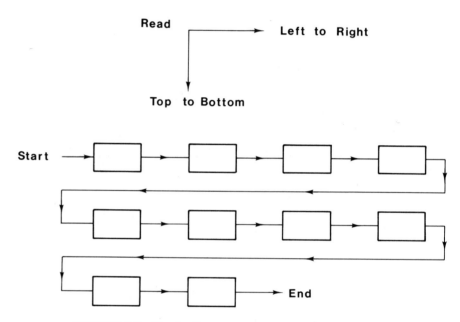

FIGURE 5-2  Block diagrams are read from left to right
and top to bottom.

Block diagrams are usually drawn on paper printed with a non-reproducible gridded pattern (Figure 5-3). This paper makes it much easier to lay out equal-sized aligned blocks, and does not require the use of a scale. If gridded drawing paper is not available, a gridded board cover may be used.

Block diagrams are used not only to express graphically the relationship between a series of elements but can also be used to analyze a series of elements. Consider the following problem. You have a coin, one side of which is a head and the other side a tail, and you are going to flip the coin four times. What are the chances of getting a head four consecutive times, and what are the chances of getting two heads and two tails in any order? Unless you are familiar with statistics, this problem can be very difficult to reason out. However, if you draw a block diagram showing the possible sequence of events, the problem can easily be solved. Figure 5-4 illustrates the four coin flips. We can see

102

**FIGURE 5-3**  Drawing paper with a nonreproducible grid.

**FIGURE 5-4**  Block diagram that illustrates four flips of a coin.

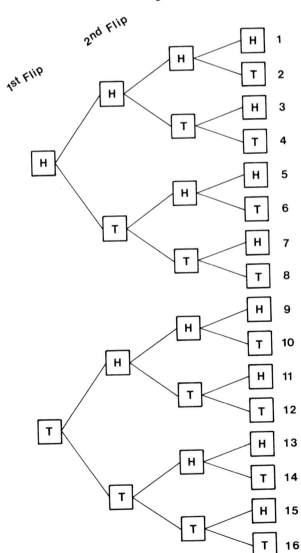

103

from Figure 5-4 that there are 16 possible combinations of heads and tails in four flips. Only one out of the 16 consists only of heads (combination 1). Therefore, the chances of flipping four consecutive heads in four flips are 1 in 16.

Combinations 4, 6, 7, 10, 11, and 13 will yield combinations of two heads and two tails in any order, so the chances of flipping two heads and two tails in any order are 6 in 16.

Figures 5-5 and 5-6 are further examples of block diagrams.

## FILTER   CIRCUIT

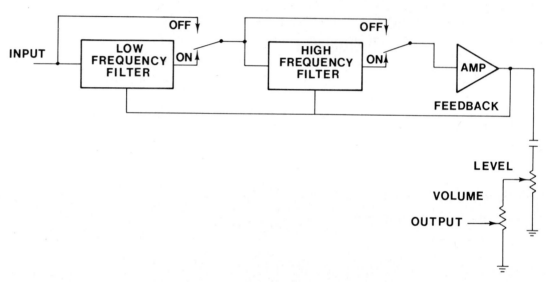

FIGURE 5-5  Example of a block diagram. (*Courtesy of General Electric Company.*)

FIGURE 5-6  Example of a block diagram. (*Courtesy of General Electric Company.*)

## RECEIVER

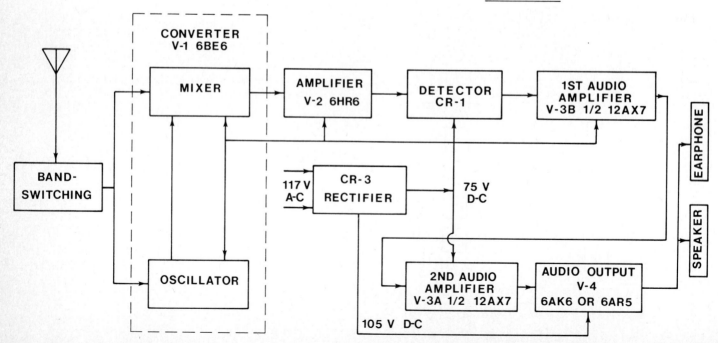

## 5-3   LOGIC DIAGRAMS

Logic diagrams are used to help design and analyze electronic circuits. They are particularly helpful in the analysis of circuits which are based on binary principles, such as those found in computers and electronic calculators.

There are several different logic functions, each of which has its own logic symbol. The logic symbols are defined in Figure 5-7 and the dimensions for some of the symbols are presented in Figure 5-8. The sizes defined in Figure 5-8 are in concurrence with the Department of Defense standard MIL-STD-806, Graphic Symbols for Logic Diagrams. These sizes are considered national standards, but may be varied as long as the basic shape remains the same.

Most drafters use templates as guides when drawing logic symbols. Figure 5-9 pictures one of the many different templates commercially available.

**AND   Function**

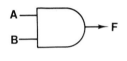

| A | B | F |
|---|---|---|
| 0 | 0 | 0 |
| 0 | 1 | 0 |
| 1 | 0 | 0 |
| 1 | 1 | 1 |

**EXCLUSIVELY  OR   Function**

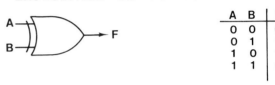

| A | B | F |
|---|---|---|
| 0 | 0 | 0 |
| 0 | 1 | 1 |
| 1 | 0 | 1 |
| 1 | 1 | 0 |

**OR   Function**

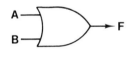

| A | B | F |
|---|---|---|
| 0 | 0 | 0 |
| 0 | 1 | 1 |
| 1 | 0 | 1 |
| 1 | 1 | 1 |

**OTHER   SYMBOLS**

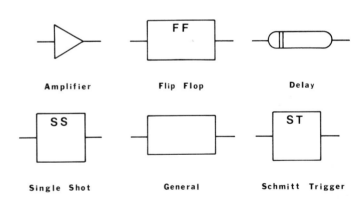

**NAND   Function   (negative AND)**

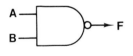

| A | B | F |
|---|---|---|
| 0 | 0 | 1 |
| 0 | 1 | 1 |
| 1 | 0 | 1 |
| 1 | 1 | 0 |

**FIGURE 5-7**  Graphic  symbols  for  logic  functions  together with their truth tables.

**NOR   Function   (negative OR )**

| A | B | F |
|---|---|---|
| 0 | 0 | 1 |
| 0 | 1 | 0 |
| 1 | 0 | 0 |
| 1 | 1 | 0 |

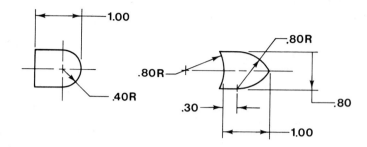

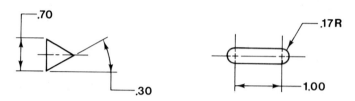

**FIGURE 5-8**   Recommended dimensions for various logic symbols.

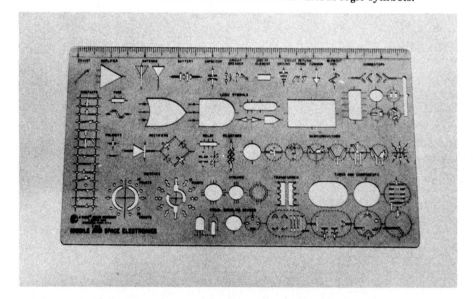

**FIGURE 5-9**   Template that includes logic symbol cutouts.

Logic diagrams are analyzed using *truth tables*. Figure 5-7 shows several basic logic functions and their truth tables. Truth tables are used to analyze not only individual logic functions but also those used in combination. For example, study the circuit shown in Figure 5-10. If the input is $A = 1, B = 0, C = 0$, the AND function will generate 0 output. When this 0 output is combined with the $C = 0$ input into the OR function, the final output will be 0. If we change the inputs to $A = 0$, $B = 0$, $C = 1$, the AND function will produce a 0 output. This 0 coupled with the $C = 1$ input will, when acted upon by the OR function, generate a final output of 1. The truth table shown in Figure 5-10 analyzes all the possible combinations for the circuit.

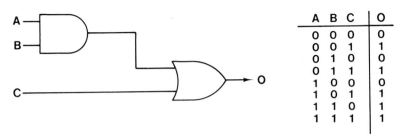

| A | B | C | O |
|---|---|---|---|
| 0 | 0 | 0 | 0 |
| 0 | 0 | 1 | 1 |
| 0 | 1 | 0 | 0 |
| 0 | 1 | 1 | 1 |
| 1 | 0 | 0 | 0 |
| 1 | 0 | 1 | 1 |
| 1 | 1 | 0 | 1 |
| 1 | 1 | 1 | 1 |

**FIGURE 5-10**  Example of a logic diagram together with its truth table.

**FIGURE 5-11**
Possible combinations for up to five inputs. Note that for two inputs there are 4 possible combinations; for three inputs, 8 combinations; for four inputs, 16 combinations; for five inputs, 32 combinations; and so on.

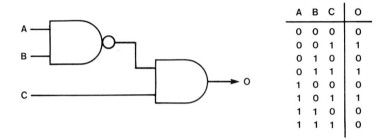

| A | B | C | O |
|---|---|---|---|
| 0 | 0 | 0 | 0 |
| 0 | 0 | 1 | 1 |
| 0 | 1 | 0 | 0 |
| 0 | 1 | 1 | 1 |
| 1 | 0 | 0 | 0 |
| 1 | 0 | 1 | 1 |
| 1 | 1 | 0 | 0 |
| 1 | 1 | 1 | 0 |

**FIGURE 5-12**  Example of a logic diagram together with its truth table.

| E | D | C | B | A | |
|---|---|---|---|---|---|
| 0 | 0 | 0 | 0 | 0 | |
| 0 | 0 | 0 | 0 | 1 | |
| 0 | 0 | 0 | 1 | 0 | 2 Inputs |
| 0 | 0 | 0 | 1 | 1 | |
| 0 | 0 | 1 | 0 | 0 | |
| 0 | 0 | 1 | 0 | 1 | |
| 0 | 0 | 1 | 1 | 0 | 3 Inputs |
| 0 | 0 | 1 | 1 | 1 | |
| 0 | 1 | 0 | 0 | 0 | |
| 0 | 1 | 0 | 0 | 1 | |
| 0 | 1 | 0 | 1 | 0 | |
| 0 | 1 | 0 | 1 | 1 | |
| 0 | 1 | 1 | 0 | 0 | |
| 0 | 1 | 1 | 0 | 1 | |
| 0 | 1 | 1 | 1 | 0 | 4 Inputs |
| 0 | 1 | 1 | 1 | 1 | |
| 1 | 0 | 0 | 0 | 0 | |
| 1 | 0 | 0 | 0 | 1 | |
| 1 | 0 | 0 | 1 | 0 | |
| 1 | 0 | 0 | 1 | 1 | |
| 1 | 0 | 1 | 0 | 0 | |
| 1 | 0 | 1 | 0 | 1 | |
| 1 | 0 | 1 | 1 | 0 | |
| 1 | 0 | 1 | 1 | 1 | |
| 1 | 1 | 0 | 0 | 0 | |
| 1 | 1 | 0 | 0 | 1 | |
| 1 | 1 | 0 | 1 | 0 | |
| 1 | 1 | 0 | 1 | 1 | |
| 1 | 1 | 1 | 0 | 0 | |
| 1 | 1 | 1 | 0 | 1 | 5 Inputs |
| 1 | 1 | 1 | 1 | 0 | |
| 1 | 1 | 1 | 1 | 1 | |

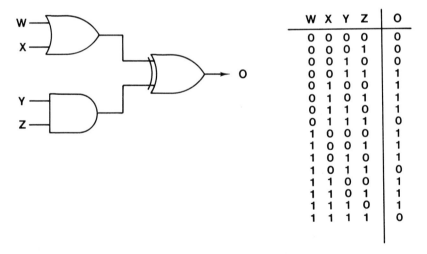

| W | X | Y | Z | O |
|---|---|---|---|---|
| 0 | 0 | 0 | 0 | 0 |
| 0 | 0 | 0 | 1 | 0 |
| 0 | 0 | 1 | 0 | 0 |
| 0 | 0 | 1 | 1 | 1 |
| 0 | 1 | 0 | 0 | 1 |
| 0 | 1 | 0 | 1 | 1 |
| 0 | 1 | 1 | 0 | 1 |
| 0 | 1 | 1 | 1 | 0 |
| 1 | 0 | 0 | 0 | 1 |
| 1 | 0 | 0 | 1 | 1 |
| 1 | 0 | 1 | 0 | 1 |
| 1 | 0 | 1 | 1 | 0 |
| 1 | 1 | 0 | 0 | 1 |
| 1 | 1 | 0 | 1 | 1 |
| 1 | 1 | 1 | 0 | 1 |
| 1 | 1 | 1 | 1 | 0 |

**FIGURE 5-13**  Example of a logic diagram together with its truth table.

   Sometimes it is difficult to remember all the possible combinations that a number of inputs could generate. Figure 5-11 has been included to help define all the possible combinations for up to five given inputs.

   Figures 5-12 and 5-13 are examples of logic diagrams together with their truth tables. Study each one and verify the meaning of each set of input combinations.

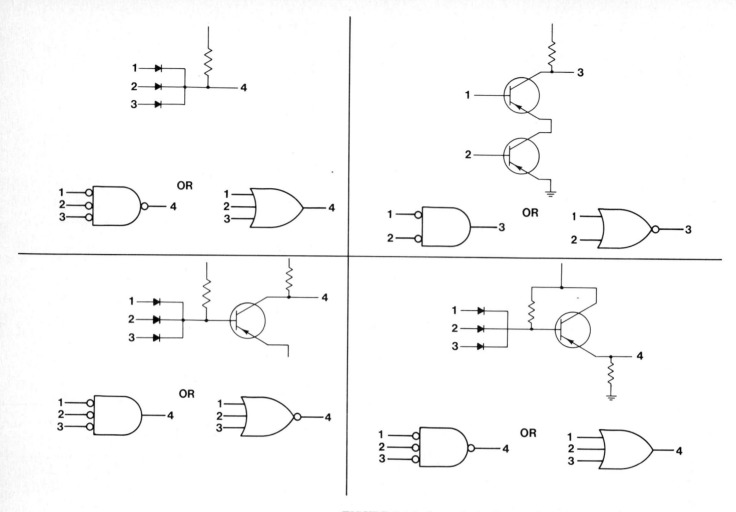

**FIGURE 5-14** Some logic diagrams with their equivalent schematic diagrams.

Figure 5-14 compares some logic diagrams with their equivalent schematic representations.

The notations 0 and 1 as used in truth tables are not numerical digits. They do *not* mean zero current or a current of unit value 1. The notations 0 and 1 represent the on and off function of a gate. They can also represent low and high voltages within a circuit. Figure 5-15 shows

**FIGURE 5-15** Different types of truth-table notations.

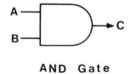

AND Gate

| IN | | OUT |
|---|---|---|
| A | B | C |
| OFF | OFF | OFF |
| OFF | ON | OFF |
| ON | OFF | ON |
| ON | ON | ON |

| IN | | OUT |
|---|---|---|
| A | B | C |
| 0 | 0 | 0 |
| 0 | 1 | 0 |
| 1 | 0 | 0 |
| 1 | 1 | 1 |

| IN | | OUT |
|---|---|---|
| A | B | C |
| L | L | L |
| L | H | L |
| H | L | L |
| H | H | H |

three different types of truth tables for an AND gate. The three tables are equivalent; only the notations are different.

Truth tables may be shortened when certain inputs are obvious or when only a certain number of inputs are possible. For example, the truth table for the logic circuit shown in Figure 5-16 has a maximum of 32 possible outputs. But if we know that input A is always 1, and inputs D and E are always 0, we can rewrite the truth table as shown.

The truth table shown (Figure 5-17) also has 32 possible outputs. This time we are interested only in outputs that generate values of 1, so we can eliminate all combinations that yield 1 and shorten the table as shown.

**FIGURE 5-16** Shortened truth table.

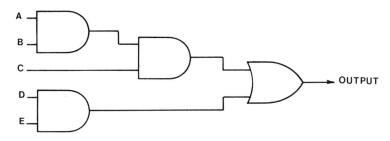

| E | D | C | B | A | OUTPUT |
|---|---|---|---|---|--------|
| 0 | 0 | 0 | 0 | 0 | 0 |
| 0 | 0 | 0 | 0 | 1 | 0 |
| 0 | 0 | 0 | 1 | 0 | 0 |
| 0 | 0 | 0 | 1 | 1 | 0 |
| 0 | 0 | 1 | 0 | 0 | 0 |
| 0 | 0 | 1 | 0 | 1 | 0 |
| 0 | 0 | 1 | 1 | 0 | 0 |
| 0 | 0 | 1 | 1 | 1 | 1 |
| 0 | 1 | 0 | 0 | 0 | 0 |
| 0 | 1 | 0 | 0 | 1 | 0 |
| 0 | 1 | 0 | 1 | 0 | 0 |
| 0 | 1 | 0 | 1 | 1 | 0 |
| 0 | 1 | 1 | 0 | 0 | 0 |
| 0 | 1 | 1 | 0 | 1 | 0 |
| 0 | 1 | 1 | 1 | 0 | 0 |
| 0 | 1 | 1 | 1 | 1 | 1 |
| 1 | 0 | 0 | 0 | 0 | 0 |
| 1 | 0 | 0 | 0 | 1 | 0 |
| 1 | 0 | 0 | 1 | 0 | 0 |
| 1 | 0 | 0 | 1 | 1 | 0 |
| 1 | 0 | 1 | 0 | 0 | 0 |
| 1 | 0 | 1 | 0 | 1 | 0 |
| 1 | 0 | 1 | 1 | 0 | 0 |
| 1 | 0 | 1 | 1 | 1 | 1 |
| 1 | 1 | 0 | 0 | 0 | 1 |
| 1 | 1 | 0 | 0 | 1 | 1 |
| 1 | 1 | 0 | 1 | 0 | 1 |
| 1 | 1 | 0 | 1 | 1 | 1 |
| 1 | 1 | 1 | 0 | 0 | 1 |
| 1 | 1 | 1 | 0 | 1 | 1 |
| 1 | 1 | 1 | 1 | 0 | 1 |
| 1 | 1 | 1 | 1 | 1 | 1 |

IF    A = 1    Reduce Table To
D = 0
E = 0

| E | D | C | B | A | OUTPUT |
|---|---|---|---|---|--------|
| 0 | 0 | 0 | 0 | 1 | 0 |
| 0 | 0 | 0 | 1 | 1 | 0 |
| 0 | 0 | 1 | 0 | 1 | 0 |

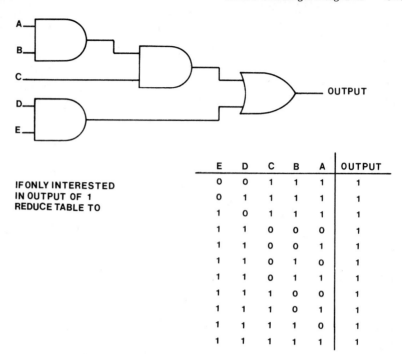

| E | D | C | B | A | OUTPUT |
|---|---|---|---|---|---|
| 0 | 0 | 1 | 1 | 1 | 1 |
| 0 | 1 | 1 | 1 | 1 | 1 |
| 1 | 0 | 1 | 1 | 1 | 1 |
| 1 | 1 | 0 | 0 | 0 | 1 |
| 1 | 1 | 0 | 0 | 1 | 1 |
| 1 | 1 | 0 | 1 | 0 | 1 |
| 1 | 1 | 0 | 1 | 1 | 1 |
| 1 | 1 | 1 | 0 | 0 | 1 |
| 1 | 1 | 1 | 0 | 1 | 1 |
| 1 | 1 | 1 | 1 | 0 | 1 |
| 1 | 1 | 1 | 1 | 1 | 1 |

IF ONLY INTERESTED
IN OUTPUT OF 1
REDUCE TABLE TO

**FIGURE 5-17**  Shortened truth table.

## 5-4  LOGIC SYMBOLS
## AND SCHEMATIC DIAGRAMS

Logic symbols are often included as part of a schematic diagram to help simplify the diagrams. The logic symbols may be included as part of an IC to help define the IC's function in relation to the rest of the schematic, as shown in Figure 5-18, or it may be used as a separate function, as shown in Figure 5-19.

Logic symbols drawn in schematic diagrams are drawn using the size dimensions outlined in Figure 5-8. Templates that meet ANSI standards may also be used. Schematics are still read from left to right, going from input to output.

**FIGURE 5-18**  Logic symbols used to explain an IC's function within a circuit.

**FIGURE 5-19**  Logic symbols used to explain the function of an IC chip.

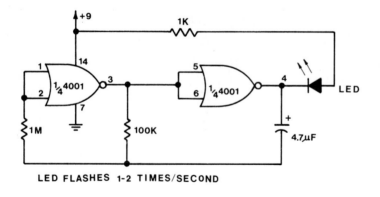

LED FLASHES 1-2 TIMES/SECOND

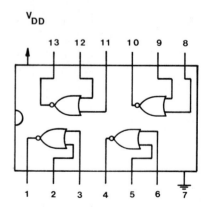

7402   QUAD   NOR   GATE

## PROBLEMS

5-1   Redraw the block diagram shown in Figure 5-1 and add two more speakers.

5-2   Prepare a block diagram that displays the following series of events.

$$\begin{array}{ll}
& \text{Get up} \\
& \text{Go to school} \\
& \text{Go to class} \\
& \text{Go to lunch} \\
\text{Go to class} & \text{Play cards} \\
\text{Go home} & \text{Play more cards} \\
\text{Study} & \text{Go home} \\
& \text{Eat supper}
\end{array}$$

5-3   Redraw, using instruments, the block diagram shown in Figure P5-3.

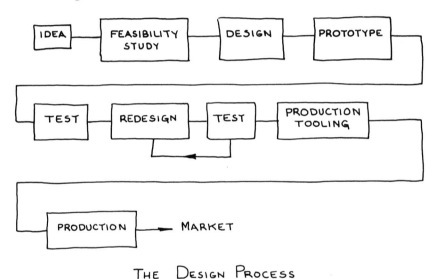

THE DESIGN PROCESS

**FIGURE P5-3**

5-4   Redraw the logic diagrams in Figure P5-4 and complete the truth tables.

**FIGURE P5-4**

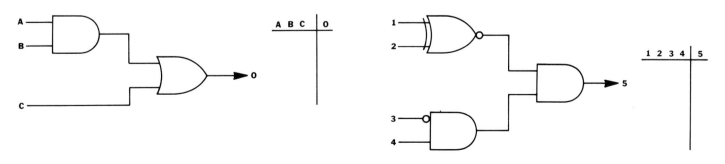

**5-5**  Redraw the following truth table and add a logic diagram whose functions will generate the truth table.

| A | B | C | Output |
|---|---|---|--------|
| 0 | 0 | 0 | 0 |
| 0 | 0 | 1 | 0 |
| 0 | 1 | 0 | 0 |
| 0 | 1 | 1 | 0 |
| 1 | 0 | 0 | 0 |
| 1 | 0 | 1 | 0 |
| 1 | 1 | 0 | 0 |
| 1 | 1 | 1 | 1 |

**5-6**  Redraw the logic diagram shown in Figure P5-6 and add the truth table.

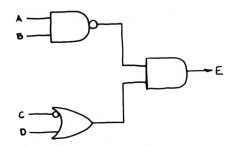

**FIGURE P5-6**

**5-7**  Redraw the logic diagrams shown in Figure P5-7 and add the truth table.

**FIGURE P5-7**

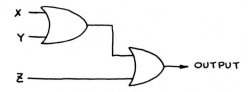

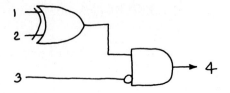

5-8   Redraw the logic diagram shown in Figure P5-8 and complete
a truth table for any 6 of the 64 possible input combinations.

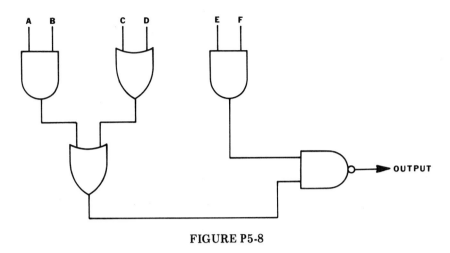

**FIGURE P5-8**

5-9     Redraw Figures P5-9 through P5-12.
**through**
5-12

**FIGURE P5-9**

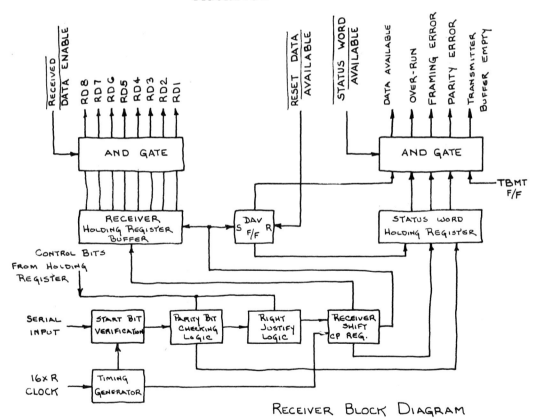

RECEIVER  BLOCK  DIAGRAM

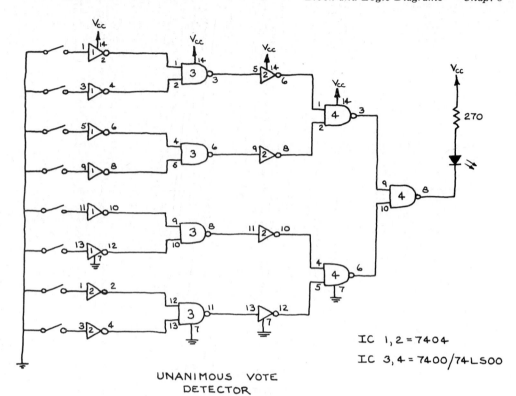

IC 1, 2 = 7404

IC 3, 4 = 7400/74LS00

UNANIMOUS VOTE
DETECTOR

**FIGURE P5-10**

**FIGURE P5-11**

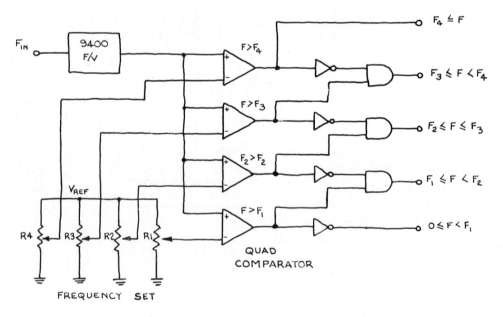

QUAD
COMPARATOR

FREQUENCY SET

FREQUENCY/TONE DECODER

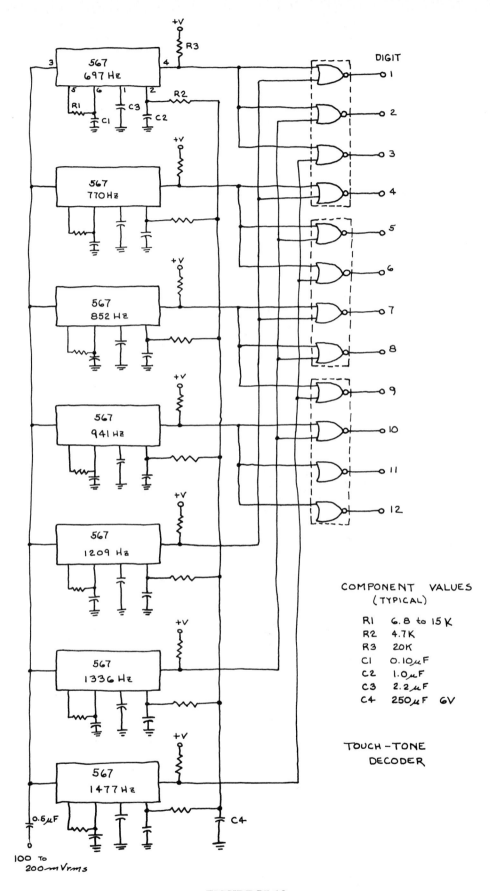

FIGURE P5-12

115

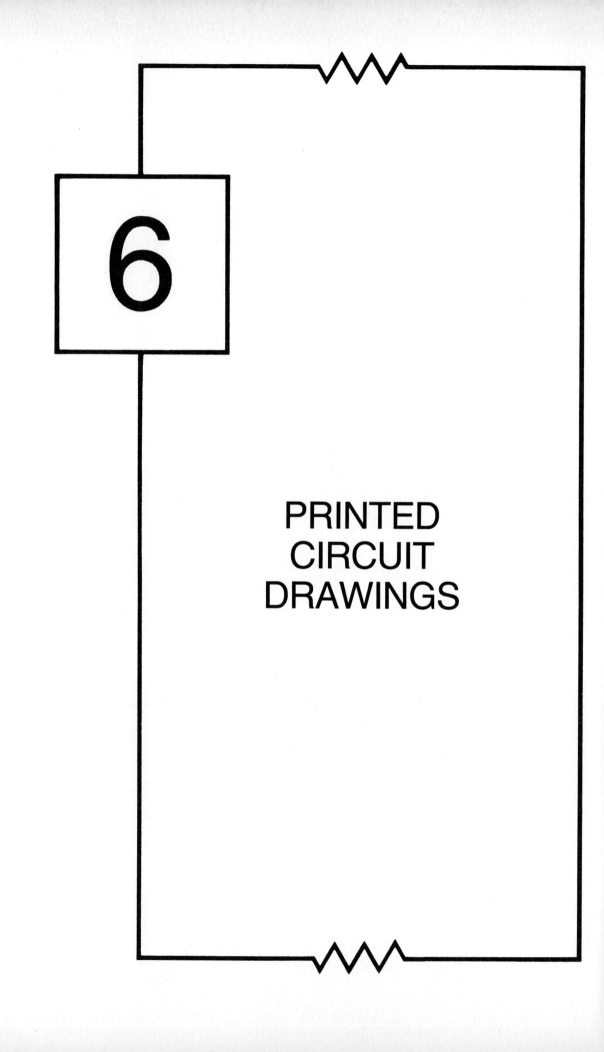

# 6

## PRINTED CIRCUIT DRAWINGS

## 6-1 INTRODUCTION

In this chapter we study how to prepare the various drawings involved in the design and manufacturing of printed circuits (PCs). These drawings include freehand design sketches, the use of paper templates, space allocation layouts, conductor paths, taping, solder masks, and drill drawings.

## 6-2 WORKING FROM SCHEMATICS

PC drawings are created from schematic diagrams such as the one shown in Figure 6-1. As we learned from Chapter 3, schematic diagrams are design drawings that show the function of a circuit. They do not necessarily have any relationship to the physical size and location of the actual components. For example, all resistors use the same symbol even though they may all have different sizes.

The first step in creating a set of PC drawings is to change the initial schematic diagram from one that uses representative symbols to one that uses pictorial symbols. Usually, this is done using a freehand sketch such as the one shown in Figure 6-2.

The pictorial schematic diagram is used to help determine the number and type (value) of the components required. Some drafters/ designers prefer to work directly from the original schematic, but sketching a pictorial schematic takes very little time and helps to maintain the functional relationship between components.

Figure 6-3 shows a schematic diagram [part (a)] and a pictorial diagram [part (b)] which utilize an IC chip. Not all the pins are needed;

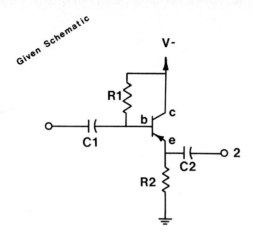

**FIGURE 6-1** Schematic diagram.

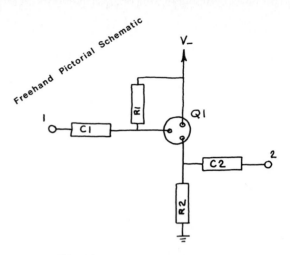

**FIGURE 6-2** Freehand pictorial sketch of the schematic diagram in Figure 6-1.

Freehand Pictorial Schematic

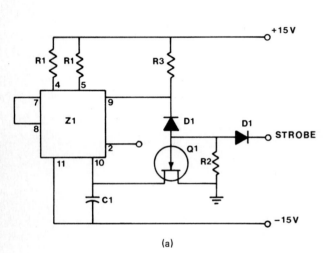

(a)

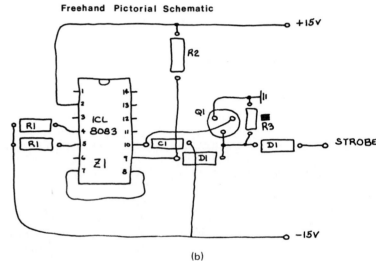

(b)

**FIGURE 6-3** (a) Schematic diagram and (b) its pictorial freehand sketch.

however, all pins are numbered. Figure 6-4(a) shows a schematic diagram in which an IC chip is drawn as two different halves. The pictorial diagram [Figure 6-4(b)] shows the chip as one unit.

## 6-3 DESIGN PRINCIPLES

After converting a schematic into a pictorial schematic, the components and conductor paths must be arranged on the board. Several design principles must be considered when arranging the components.

If possible, all components would be aligned in a common direction, as shown in Figure 6-5. Uniform alignment helps simplify PC board manufacture. Keep conductor paths as short as possible, and avoid sharp turns and acute angles, as shown in Figure 6-6.

If possible, make all conductor path intersections on pads, as shown in Figure 6-7. Intersections made directly into conductor paths tend to develop microscopic cracks and thereby cause poor reliability.

Conductor paths should be arranged to minimize crossovers, as shown in Figure 6-8. Conductor paths on PC boards are not shielded so

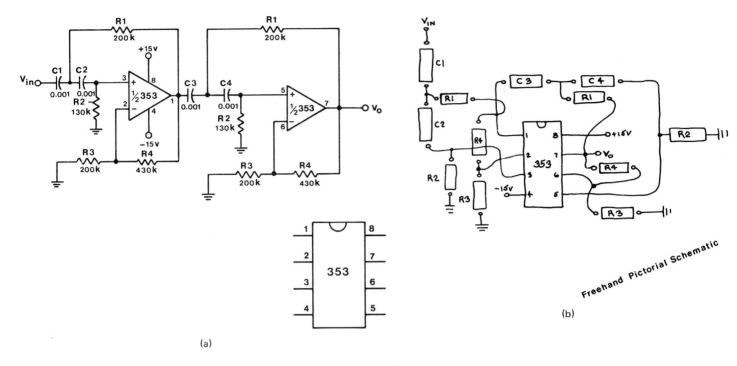

(a)

**FIGURE 6-4** (a) Schematic diagram and (b) its pictorial freehand sketch.

**FIGURE 6-5** How to lay out a PC diagram.

**Best    arrangement**

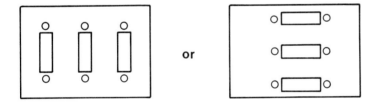

or

**An  acceptable  arrangement**

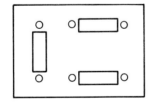

**Avoid !!**

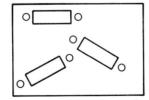

## CONDUCTOR PATHS

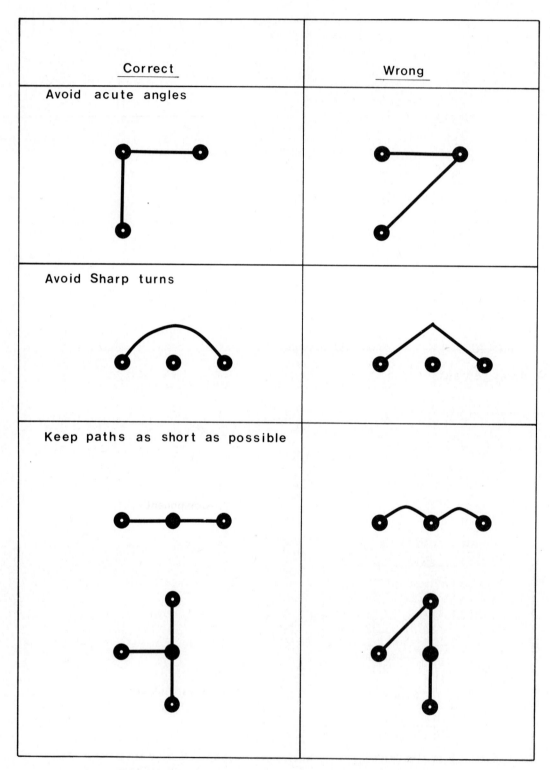

**FIGURE 6-6** Keep all conductor paths as short as possible and avoid sharp turns.

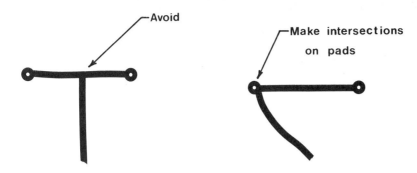

FIGURE 6-7  Avoid T connections.

FIGURE 6-8  Avoid crossovers.

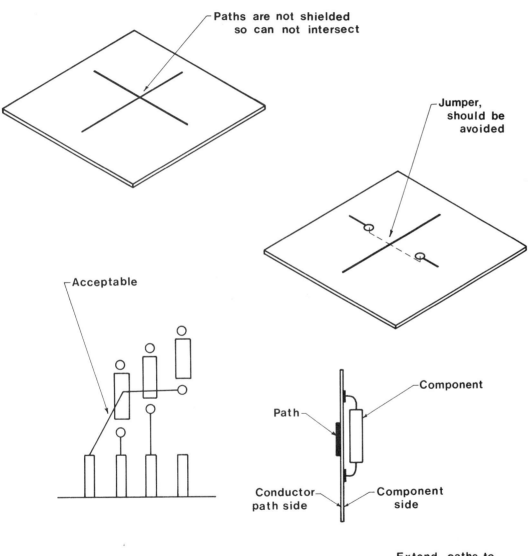

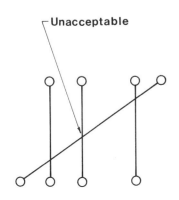

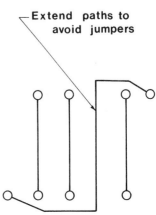

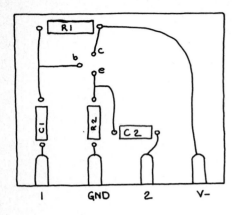

1ST ATTEMPT

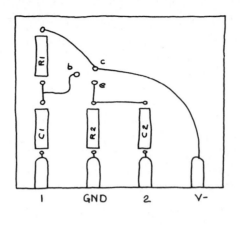

2ND ATTEMPT

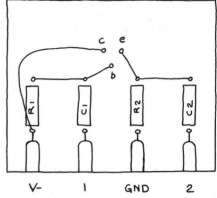

3RD ATTEMPT

**FIGURE 6-9**   Arranging a PC board layout.

**FIGURE 6-10**  Freehand design sketches of a component arrangement.

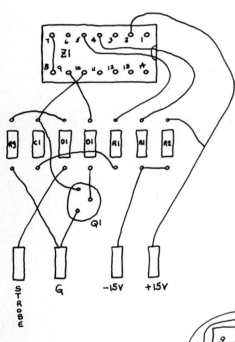

**FIGURE 6-11**
Freehand design sketches of a component arrangement.

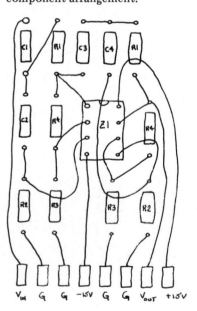

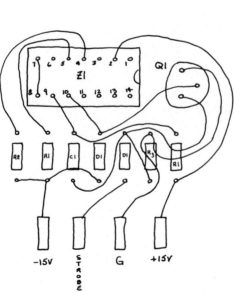

cannot crossover each other as can insulated wires. When conductor paths intersect, one of the paths must pass through the board to the opposite side, pass under the other conductor path, then return to its original side and resume its path. This procedure, called a "jumper," adds time and expense to the PC board's manufacture and should be avoided.

Figure 6-9 shows how these principles are applied to the schematic diagram presented in Figure 6-2. All inputs and outputs were located on the same board edge; then several attempts were made to align the components, minimize conductor paths, avoid jumpers, and so on. Note how different the third and final arrangement is from the original pictorial schematic.

The component rearrangement process shown in Figure 6-9 was done using freehand sketches. No attempt was made to use actual component sizes. Freehand sketching is fast and if done on graph paper can be relatively accurate. Figures 6-10 and 6-11 show similar sketches for the schematics shown in Figures 6-3 and 6-4.

## 6-4   SPACE ALLOCATION LAYOUTS (MASTER LAYOUTS)

After an acceptable component configuration has been sketched, the size of components is considered. It should be pointed out that after the component sizes are considered, the sketches must sometimes be reworked. Component sizes may vary considerably, so a back-and-forth process between arrangement sketches and space allocation layout is not unusual.

The actual size of components can be found in manufacturers' catalogs. Figure 6-12 shows a sample page from a Mallory catalog. Note that the page includes information as to the components' performance and capacity as well as size data. Appendix H lists some other component sizes. Component sizes may be listed in either English or metric units.

The mounting requirement of each component is also an important consideration. Space must be alloted not just for the component, but also for the area needed to mount the component to the board. For example, a resistor is mounted using wires that protrude from each side, which means that mounting size is larger than the resistor size.

The distance between the holes used to mount resistors (and cylindrically shaped capacitors and diodes) may be determined by the formula (see Figure 6-13)

$$MD = R + 8t$$

where MD = mounting distance
$R$ = component length
$t$ = wire thickness

However, almost all components have some "play" in their mounting distances. This means that the distances can be varied—made a little longer or shorter—as required to match a given grid pattern. If, for example, the mounting distance were calculated to be 0.72 inch, an eight-to-the-inch pattern, the centers of the mounting holes would be

## Polyester, Axial Leaded

M192P capacitors are wound with a polyester film dielectric and thin gage foil under carefully controlled conditions, permitting reliable working voltages of 80, 200, and 400 volts. The capacitor features extended foil sections which are terminated in metal end caps, assuring a consistent capacitor size as well as a non-inductive connection. The end caps further protect against entry of moisture into the capacitor. Physical sizes of the M192P capacitor have been chosen as closely as possible to those of various composition resistors for compatibility with automatic insertion equipment. **Request bulletin 9-781 for complete specifications. For pricing, reference price sheet No. 430.**

**Highlights**
Capacitance: .0001μF to .39μF
Voltage: 80, 200 and 400 WVDC
Tolerance: ±10%
Temperature: -55°C to +85°C

### 80 WVDC

| Capacitance μF | Diameter | Length | Catalog Number |
|---|---|---|---|
| .0022 | 0.138 | 0.312 | M192P2229R8 |
| .0027 | 0.138 | 0.312 | M192P2729R8 |
| .0033 | 0.138 | 0.312 | M192P3329R8 |
| .0039 | 0.138 | 0.312 | M192P3929R8 |
| .0047 | 0.138 | 0.312 | M192P4729R8 |
| .0056 | 0.170 | 0.375 | M192P5629R8 |
| .0068 | 0.170 | 0.375 | M192P6829R8 |
| .0082 | 0.170 | 0.375 | M192P8229R8 |
| .01 | 0.170 | 0.375 | M192P1039R8 |
| .012 | 0.170 | 0.437 | M192P1239R8 |
| .015 | 0.170 | 0.437 | M192P1539R8 |
| .018 | 0.170 | 0.500 | M192P1839R8 |
| .022 | 0.170 | 0.500 | M192P2239R8 |
| .027 | 0.204 | 0.437 | M192P2739R8 |
| .033 | 0.204 | 0.437 | M192P3339R8 |
| .039 | 0.204 | 0.562 | M192P3939R8 |
| .047 | 0.204 | 0.562 | M192P4739R8 |
| .056 | 0.225 | 0.562 | M192P5639R8 |
| .068 | 0.225 | 0.562 | M192P6839R8 |
| .082 | 0.290 | 0.500 | M192P8239R8 |
| .1 | 0.290 | 0.500 | M192P1049R8 |
| .12 | 0.290 | 0.625 | M192P1249R8 |
| .15 | 0.290 | 0.625 | M192P1549R8 |
| .22 | 0.290 | 0.625 | M192P2249R8 |
| .27 | 0.318 | 0.875 | M192P2749R8 |
| .33 | 0.318 | 0.875 | M192P3349R8 |
| .39 | 0.318 | 1.187 | M192P3949R8 |

### 200 WVDC

| Capacitance μF | Diameter | Length | Catalog Number |
|---|---|---|---|
| .0001 | 0.138 | 0.312 | M192P10192 |
| .00012 | 0.138 | 0.312 | M192P12192 |
| .00015 | 0.138 | 0.312 | M192P15192 |
| .00018 | 0.138 | 0.312 | M192P18192 |
| .00022 | 0.138 | 0.312 | M192P22192 |
| .00027 | 0.138 | 0.312 | M192P27192 |
| .00033 | 0.138 | 0.312 | M192P33192 |
| .00039 | 0.138 | 0.312 | M192P39192 |
| .00047 | 0.138 | 0.312 | M192P47192 |
| .00056 | 0.138 | 0.312 | M192P56192 |
| .00068 | 0.138 | 0.312 | M192P68192 |
| .00082 | 0.138 | 0.312 | M192P82192 |
| .001 | 0.138 | 0.312 | M192P10292 |
| .0012 | 0.138 | 0.312 | M192P12292 |
| .0015 | 0.138 | 0.312 | M192P15292 |
| .0018 | 0.170 | 0.312 | M192P18292 |
| .0022 | 0.170 | 0.312 | M192P22292 |
| .0027 | 0.170 | 0.312 | M192P27292 |
| .0033 | 0.170 | 0.312 | M192P33292 |
| .0039 | 0.170 | 0.437 | M192P39292 |
| .0047 | 0.170 | 0.437 | M192P47292 |
| .0056 | 0.170 | 0.437 | M192P56292 |
| .0068 | 0.170 | 0.437 | M192P68292 |
| .0082 | 0.170 | 0.500 | M192P82292 |
| .01 | 0.170 | 0.500 | M192P10392 |
| .012 | 0.204 | 0.437 | M192P12392 |
| .015 | 0.204 | 0.437 | M192P15392 |

### 400 WVDC

| Capacitance μF | Diameter | Length | Catalog Number |
|---|---|---|---|
| .00047 | 0.170 | 0.375 | M192P47194 |
| .00056 | 0.170 | 0.375 | M192P56194 |
| .00068 | 0.170 | 0.375 | M192P68194 |
| .00082 | 0.170 | 0.375 | M192P82194 |
| .001 | 0.170 | 0.375 | M192P10294 |
| .0012 | 0.170 | 0.375 | M192P12294 |
| .0015 | 0.170 | 0.375 | M192P15294 |
| .0018 | 0.170 | 0.437 | M192P18294 |
| .0022 | 0.170 | 0.437 | M192P22294 |
| .0027 | 0.204 | 0.437 | M192P27294 |
| .0033 | 0.204 | 0.437 | M192P33294 |
| .0039 | 0.204 | 0.437 | M192P39294 |
| .0047 | 0.204 | 0.437 | M192P47294 |
| .0056 | 0.225 | 0.562 | M192P56294 |
| .0068 | 0.225 | 0.562 | M192P68294 |
| .0082 | 0.225 | 0.562 | M192P82294 |
| .01 | 0.225 | 0.562 | M192P10394 |
| .012 | 0.290 | 0.500 | M192P12394 |
| .015 | 0.290 | 0.500 | M192P15394 |
| .018 | 0.290 | 0.625 | M192P18394 |
| .022 | 0.290 | 0.625 | M192P22394 |
| .027 | 0.318 | 0.625 | M192P27394 |
| .033 | 0.318 | 0.625 | M192P33394 |
| .039 | 0.318 | 0.875 | M192P39394 |
| .047 | 0.318 | 0.875 | M192P47394 |
| .056 | 0.318 | 1.187 | M192P56394 |
| .068 | 0.318 | 1.187 | M192P68394 |

**FIGURE 6-12** Sample manufacturer's catalog. *(Courtesy of Mallory.)*

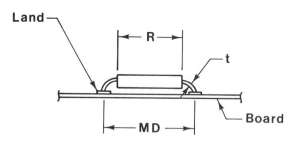

**FIGURE 6-13** How to determine the distance between resistor mounting holes.

located 0.75 inch apart. On a ten-to-the-inch pattern, it would be 0.70 inch. Figure 6-14 shows this concept.

Component space allocation layouts are usually prepared to match a given grid pattern. Eight-to-the-inch, ten-to-the-inch, and metric patterns are the most popular. Ten-to-the-inch patterns have become increasingly popular, as IC chips have mounting pins spaced 0.10 inch apart. The distance 0.10 inch is almost equal to 2.5 mm, so chips made to this mounting distance can easily be understood by designers and manufacturers who work in the metric system (see Figure 6-15).

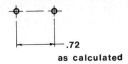
.72
as calculated

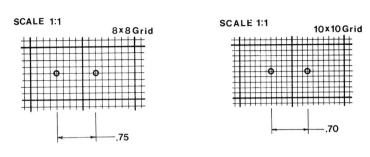
SCALE 1:1    8×8 Grid

.75

SCALE 1:1    10×10 Grid

.70

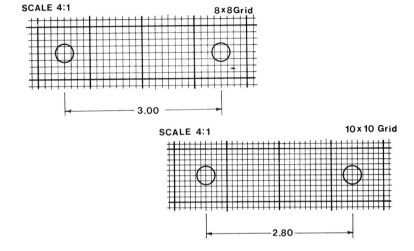
SCALE 4:1    8×8 Grid

3.00

SCALE 4:1    10×10 Grid

2.80

FIGURE 6-14   Layout should match the grid pattern.

FIGURE 6-15   Layout should match the grid pattern.

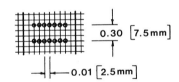
Scale 1:1

0.30 [7.5mm]

0.01 [2.5mm]

ALWAYS TRY TO MATCH THE GRID PATTERN

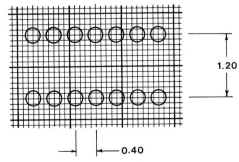
Scale 4:1

1.20

0.40

125

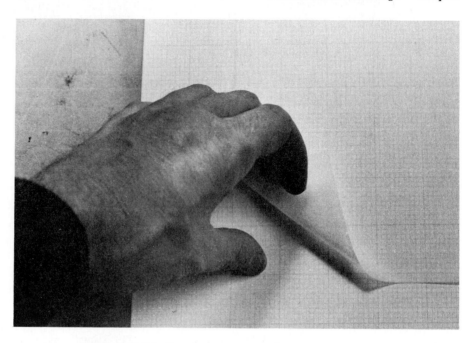

**FIGURE 6-16**  Drawing paper taped over a sheet of graph paper.

Space allocation layouts incorporate grid patterns in one of two ways. A sheet of graph paper may be taped to the drawing board, then a clear (transparent) sheet of drawing paper taped over the graph paper, as shown in Figure 6-16. The grid pattern can then be seen through the clear paper. Further examples of this technique are shown in Figures 1-3 through 1-6.

Figure 6-17 shows a sheet of drawing paper that has a grid pattern printed on it. The grid is printed using nonreproducible blue ink so that the grid will not appear in blueprints of the drawings.

**FIGURE 6-17**  Sheet of drawing paper with a preprinted grid pattern.

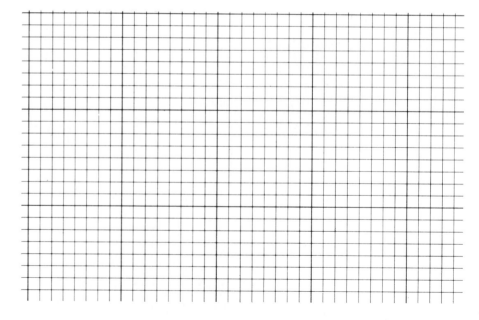

Space allocation layouts are usually prepared four times larger than actual size. This is referred to as drawing "4 up" or "4 times scale." The drawing scale is defined as 4:1.

An increased scale is used to help ensure drawing, and therefore manufacturing accuracy. Any errors will be only one-fourth of their actual size when the drawing is reduced to final size. Layouts can be done to other scales, but 4:1 is the most popular.

After a grid pattern and scale have been chosen and affixed to the drawing board, the next step is to define the component border, the PC board size, and the grid locators, as shown in Figure 6-18.

The *component border* is the area of the PC board that can be used to mount components. This area differs from the overall PC board area because it does not include any areas needed for mounting the board or which would interfere with other devices or boards. The component border is clearly defined by a heavy black line.

The *PC board size* is defined by using corner brackets. The corner brackets are heavy black lines drawn perpendicular to each other and to each corner of the board. The inside of the bracket should match the outside of the board. A single linear dimension is included on the layout together with the note "REDUCE TO . . ." to define the final board size. A note such as "REDUCE TO 2.00 ± .01" would mean that during the final photographic processes in manufacturing, this dimension would be reduced until it exactly matched the specified 2.00 ± .01 inches.

**FIGURE 6-18**　PC board layout.

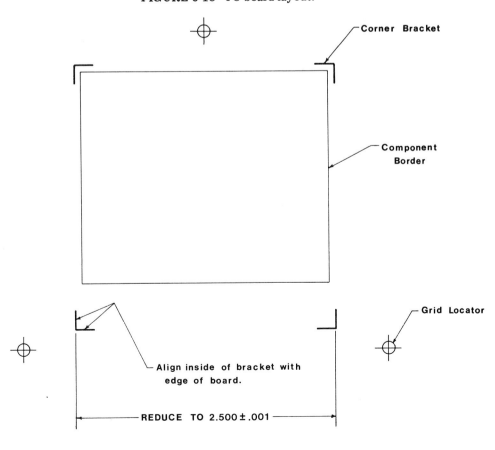

*Grid locators* are symbols placed on the drawing to identify the location of the grid and to be used to help align the space allocation layout with related PC drawings. The crossed perpendicular lines must align exactly with the lines of the grid pattern. The circle is approximately 0.25 inch in diameter.

The components are located within the component border using one of two techniques: drawing templates or paper templates. Figure 6-19 shows an example of a component layout drawing template. Note that both the overall size and mounting hole spacing are included. Layout templates are used, as are symbol templates, which were explained in Section 2-2.

Paper templates are pieces of paper that show the size and mounting hole requirements. They are cut to size slightly larger than the component. Paper templates can be purchased or made as needed.

Many drafters prepare a master component sheet such as the one shown in Figure 6-20, then make copies as necessary. (The choice of components depends on individual requirements.) Remember that Xerox copies are approximately 2% larger than their originals, so never make copies of copies, only originals.

Figure 6-21 shows a space allocation layout for the schematic diagram presented in Figures 6-2 and 6-9. Figure 6-21(a) was prepared using a drawing template; Figure 6-21(b) was prepared using paper templates. Both were prepared using a scale of 4:1. This means that the resistors, which actually measure 0.30 inch long, were drawn 1.20 inches long. All other components were drawn in a similar manner.

**FIGURE 6-19**   Component layout template.

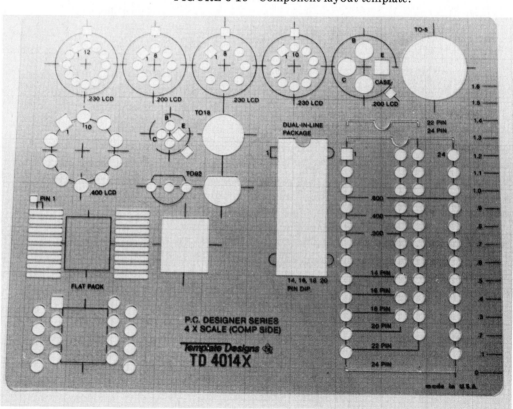

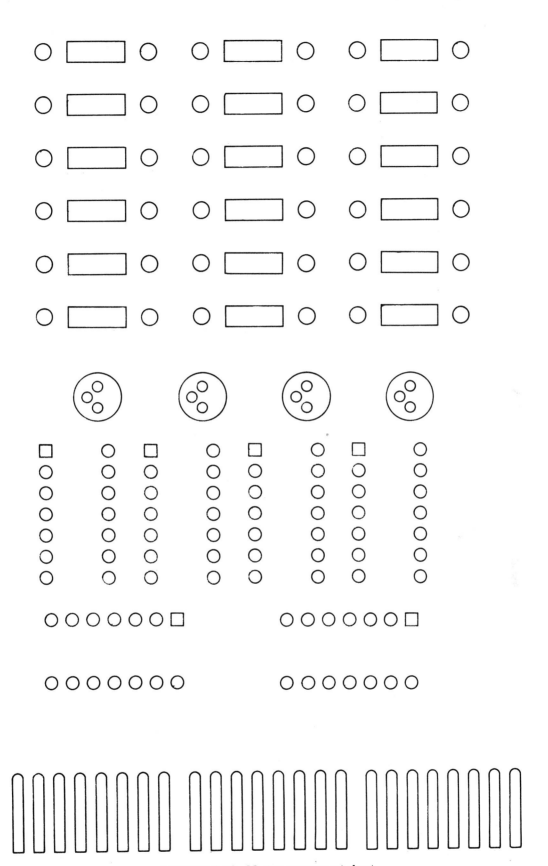

**FIGURE 6-20** Master component sheet.

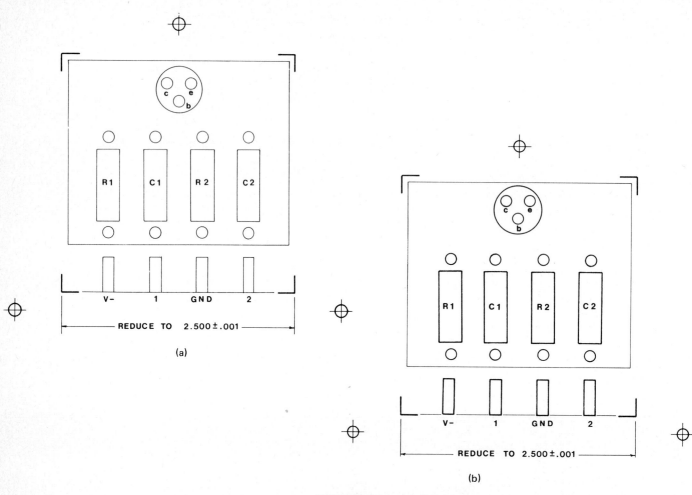

(a)

(b)

**FIGURE 6-21** (a) Space allocation layout; (b) space allocation using paper templates.

**FIGURE 6-22** Layouts are drawn as if viewed from the component side of the board.

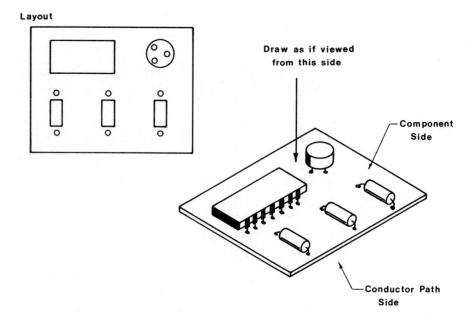

The components are always located on the top of the board; that is, the layout looks directly down on the components. The conductor paths are on the other side (the side not directly seen) of the board. Figure 6-22 shows this concept.

Figures 6-23 and 6-24 are further examples of space allocation layouts based on the schematic diagrams shown in Figures 6-3 and 6-4. They have also been drawn at a scale of 4:1.

Some drafters combine the paper template and design sketching techniques and procedures outlined in Section 6-3. After an initial rough pictorial sketch is made, a space allocation layout is prepared based on the sketch. The paper components are then moved around to try to find an acceptable pattern (no jumpers, etc.). The new pattern is then copied and checked by adding conductor paths. The procedure is repeated until an acceptable pattern is found.

**FIGURE 6-23**   Space allocation layout.

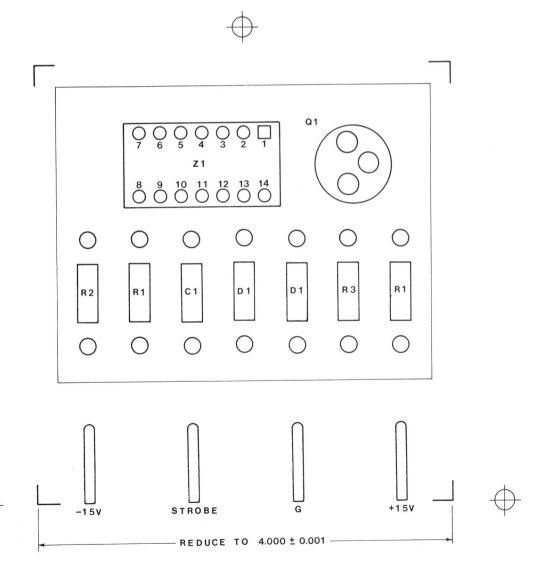

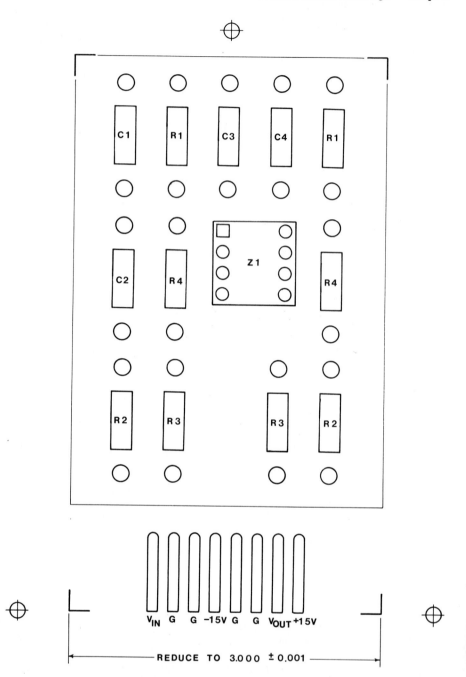

**FIGURE 6-24**  Space allocation layout.

## 6-5  CONDUCTOR PATHS

Conductor paths are added to space allocations using solid lines, as shown in Figure 6-25. The paths are usually drawn using a series of straight lines (not curved lines), as shown.

Conductor paths are usually $1/16$ (0.0625) inch wide with a minimum of $1/32$ (0.031) inch between the paths, as shown in Figure 6-26.

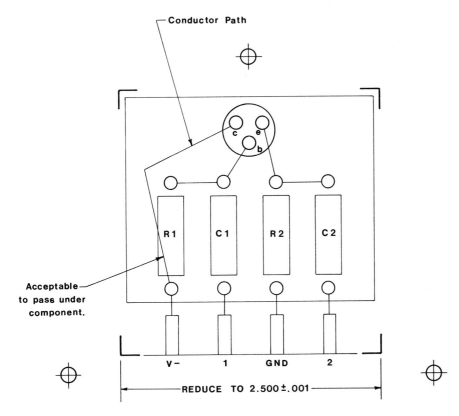

**FIGURE 6-25** Space allocation layout with conductor paths.

For very lower power ¹⁄₃₂-inch widths may be used and widths greater than ¹⁄₁₆ inch are used for large power requirements.

The correct line spacing for conductor paths on a 4:1 scaled drawing is found by the formula

$$D = 4W + 4S$$

where $D$ = distance between lines
$\qquad W$ = conductor path width
$\qquad S$ = width of space between conductor paths

This means that if the conductor path is ¹⁄₁₆ inch wide and the spacing between paths is ¹⁄₃₂ inch:

$$D = 4(¹⁄₁₆) + 4(¹⁄₃₂)$$
$$= ¹⁄₄ + ¹⁄₈$$
$$= ³⁄₈ \text{ inch}$$

Thus all lines that represent conductor paths should be drawn at least ³⁄₈ inch apart.

Figures 6-27 and 6-28 are completed space allocation layouts with conductor paths added for the schematics presented in Figures 6-3 and 6-4.

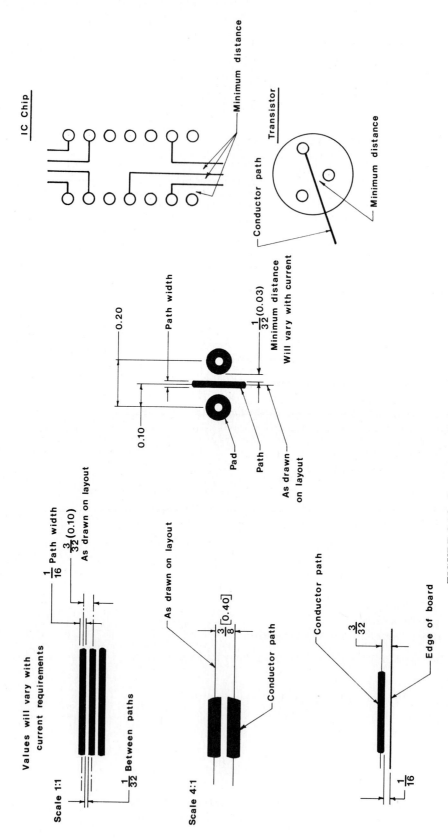

FIGURE 6-26  Layout requirements for conductor paths.

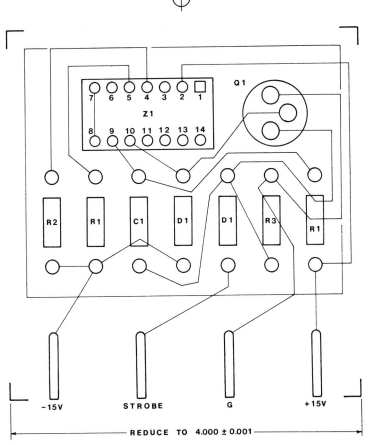

**FIGURE 6-27** Sample space allocation layout based on the schematic shown in Figure 6-3.

**FIGURE 6-28** Sample space allocation layout based on the schematic shown in Figure 6-4.

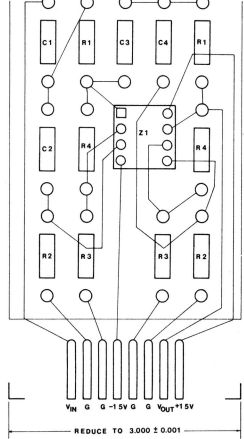

## 6-6   TAPE MASTERS

Tape masters are drawings that show all conductor paths and intersection pads. They are used in the manufacturing process for photographic definition of etching requirements. Tape masters may be thought of as manufacturing templates (see Figure 6-29).

Tape is used rather than shaded pencil lines to ensure dense lines that will reproduce clearly and accurately when photographed. Tape is also easy to peel off the drawing and move if an error in location is made.

Tape for PC tape masters is available in many different shapes and sizes. Figure 6-30 shows some of the many precut tapes that are commercially available.

Figure 6-31 shows the equipment needed to prepare a tape master. This includes the tape, both precut pads and ⅛-inch wide tape (the tape master is also done at a 4:1 scale) in rolled form; an Exacto knife for cutting tape, a brandishing pen to press the tape into place; and several samples of drawing film.

Tape masters are usually prepared on drawing film. Drafting film is a strong, durable, clear plastic film which will not discolor with age. It is available in precut standard drawing sizes or in rolled form. Figure 6-32 shows tape being applied to drawing film.

To prepare a tape master, first tape the space allocation layout (master layout) on the drawing board, then tape a sheet of drawing film over the layout as shown in Figure 6-33. The corner brackets and grid locators are then traced using tape. Next, all pads and connector strips

**FIGURE 6-29**   Tape drawing.

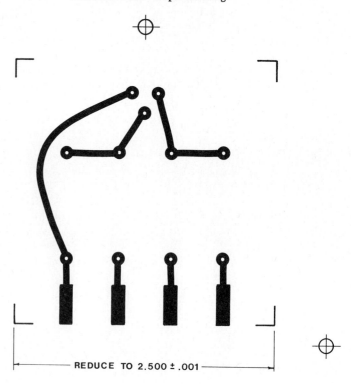

REDUCE TO 2.500 ± .001

**FIGURE 6-30**   Precut tape.

**FIGURE 6-31**   Precut tape for help in creating PC board drawings.

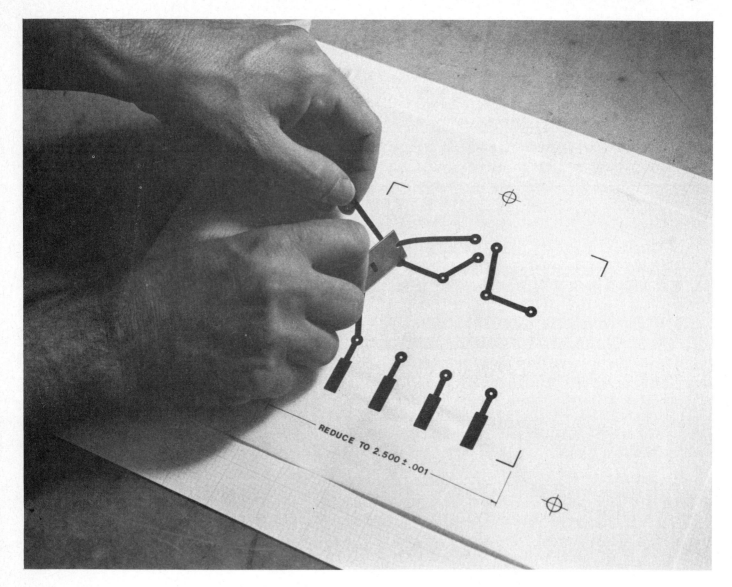

**FIGURE 6-32**  How to apply precut tape to create a PC drawing.

(fingers) are added. Conductor paths are taped after the pads and connector paths. Finally, the sizing dimension is added.

Note that the conductor paths taped on in Figure 6-34 do not
exactly match the straight conductor paths drawn on the layout. Tape
should be mounted in smooth, rounded patterns. Square edges should
be avoided, as they could lead to cracks in the paths. Rounded patterns
are also better for the actual conductor paths, as square shapes tend to
crack here also. Even T-joints, as shown in Figure 6-35, should be
reinforced to prevent cracks.

Note also that all pads maintain an open, clear center portion. This
open area is helpful when dulling the PC board and should not be filled

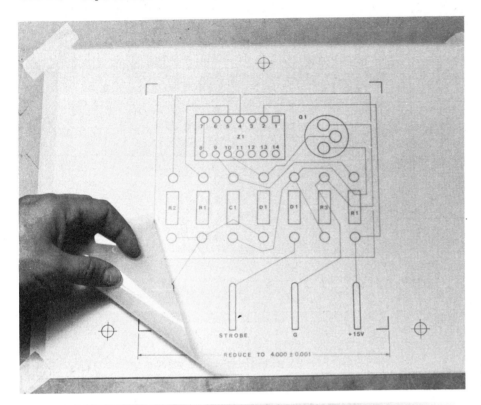

**FIGURE 6-33** Sheet of drawing film taped over a space allocation layout.

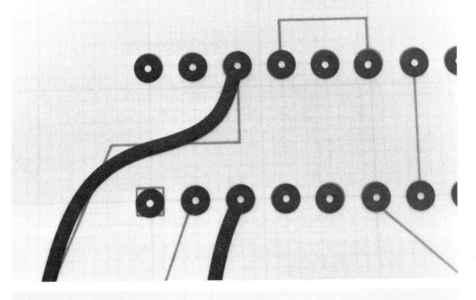

**FIGURE 6-34** Tape does not have to exactly follow the conductor paths drawn in the master layout.

**FIGURE 6-35** Precut tape T joints.

REDUCE TO 4.000±0.001

**FIGURE 6-36** Tape master for the schematic shown in Figure 6-3.

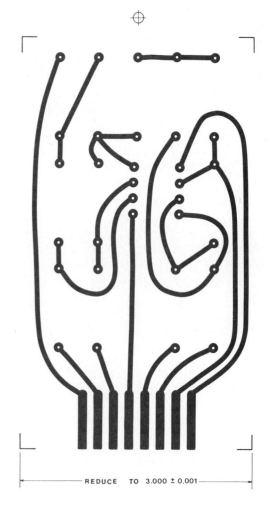

REDUCE TO 3.000 ± 0.001

**FIGURE 6-37** Tape master for the schematic shown in Figure 6-4.

**FIGURE 6-38**   Light tables.

in or omitted. Figures 6-36 and 6-37 show the tape masters for the schematics presented in Figures 6-3 and 6-4.

Figure 6-38 shows a light table. A light table is similar to a drawing table except that its top surface is made of frosted glass. Under the glass is a set of lights. Light tables are excellent for tracing because light comes from under the drawing being traced. Most professional drafters use light tables when preparing tape masters from master layouts.*

## 6-7   SOLDERING MASKS AND DRILL DRAWINGS

Soldering masks and drill drawings are special types of drawings which are not always prepared as part of a PC drawing package. However, they are very useful in automatic production work.

A soldering mask is a taped mask that shows only the pad areas of the circuit. It is prepared on drawing film and includes a solid shape (no clear area in the center) for each pad as necessary, corner brackets, grid locators, and one sizing dimension. Figure 6-39 shows examples of soldering masks. Soldering masks are copied from either tape masters or space allocation layouts. They are also done at a 4:1 scale.

Drill drawings are drawings that define the pattern of holes which must be drilled into the board using an *x-y* coordinate system. Figure 6-40 shows an example.

The hole pattern is traced from the space allocation layout. A zero reference point is chosen, which may be the corner of the board or the center of a locating hole previously drilled into the board. The *x*

---

*Students are encouraged to build their own light "boxes" using several short fluorescent light units. A working area of $12 \times 18$ inches or $18 \times 24$ inches will be acceptable for most layouts. A design for a simple light box is included in Appendix I.

**FIGURE 6-39**
Soldering mask.

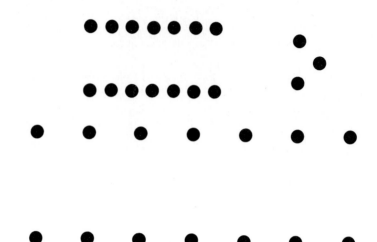

REDUCE TO 4.000 ± 0.001

**FIGURE 6-40**
Drill drawing.

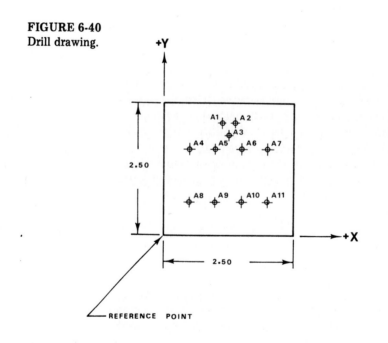

| HOLE | X | Y | DIA |
|------|------|------|------|
| A1 | 1.12 | 2.25 | .032 |
| A2 | 1.38 | 2.25 | ↑ |
| A3 | 1.25 | 1.88 | |
| A4 | 0.50 | 1.62 | |
| A5 | 1.00 | ↑ | |
| A6 | 1.50 | ↓ | |
| A7 | 2.00 | 1.62 | |
| A8 | 0.50 | 0.62 | |
| A9 | 1.00 | ↑ | |
| A10 | 1.50 | ↓ | ↓ |
| A11 | 2.50 | 0.62 | .032 |

142

and $y$ directions are defined from the zero reference point, and the minimum number of dimensions needed to size the board are added.

All hole dimensions are defined in terms of the $x$-$y$ coordinate system and presented in a chart form as shown. Note that the drawing is a trace of a drawing done at 4:1 scale; all dimensions listed in the chart must be actual size. The chart dimensions are final dimensions, that is, the size to be used in manufacturing. Divide the traced dimensions by 4 to get the actual dimensions.

To prevent error, some drafters prefer to redraw the hole pattern at a scale of 1:1 so that measurements may be taken directly. However, it is a fairly simple process to divide all measured values by 4, especially if a calculator is available.

## 6-8    DOUBLE-SIDED PC BOARDS

Double-sided PC boards differ from single-sided boards in that they have conductor paths on both sides of the board. Components are still usually restricted to one side of the board but could be mounted on both sides. Double-sided boards are more expensive and time consuming to manufacture but use a smaller area than that of single-sided boards.

Drawings for double-sided boards are prepared using the same procedure as that used for single-sided boards. There is one space allocation layout (master layout), but tape masters must be prepared for each side of the board. The increased number of drawings requires:

1. Accurate grid locators
2. Careful identification of each drawing

It is vital that the individual drawings match each other. Any drawing errors will result in manufacturing errors.

Figure 6-41 shows a space allocation layout developed from a schematic. The procedure used to locate the components and conductor paths is as explained previously, but some differences are required to distinguish which paths are on which side of the board.

The board is still viewed from the component side. The component side is considered to be the top of the board. Conductor paths on the top side (component side) use a solid line; conductor paths on the bottom side (conductor path side) use hidden lines. Some drafters use colors in addition to the line pattern difference to distinguish board sizes. Tape masters are prepared for each side of the board using the common master layout. Drawing film is used and, if available, a light table.

There are several different techniques used to identify the tape masters for different sizes of a double-sided board. One technique, shown in Figure 6-42, uses one drawing with different-colored transparent tapes. The component side is prepared using red tape, the conductor path side, using blue tape. Both sides of the drawing film are used, so when viewed from one angle the lettering on the opposite side will be reversed. During manufacturing, special camera filters are used to block out the unwanted color tape pattern.

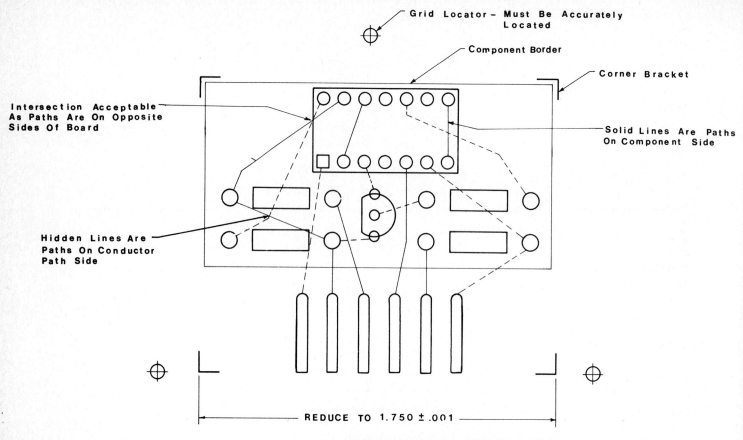

**FIGURE 6-41** Space allocation drawing for a double-sided PC board.

**FIGURE 6-42** Identifying a double-sided PC board.

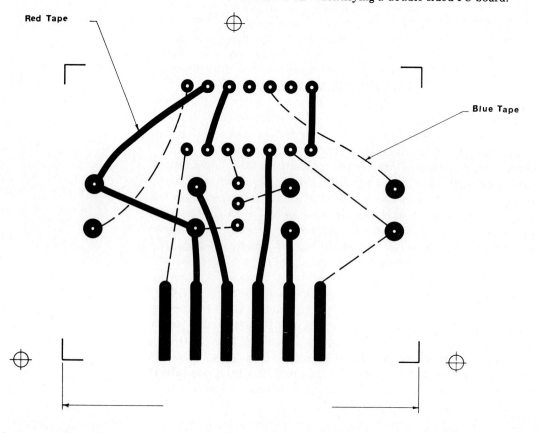

Another technique, shown in Figure 6-43, uses two separate tape masters. One defines the conductor path side pattern, the other, the component side pattern. Sometimes the lettering on the conductor path is done backward to help call attention to the fact that it is opposite the conductor path side.

A third technique, shown in Figure 6-44, uses one large sheet of drawing film. Both sides are drawn on the same sheet, separated by an end view.

As with single-sided boards, pad masters are sometimes prepared for production work. Pad masters ensure more accurate production by eliminating any mismatch between pads drawn on opposite sides of the board.

If a pad master is used, three tape masters will be required: the pad master, the conductor paths on the component side, and the conductor paths on the conductor side. Figure 6-45 shows a set of tape masters that includes a pad master. Conductor paths are always overlapped onto the pad areas.

## 6-9   COMPONENT OUTLINE DRAWINGS

Component outline drawings are drawings that outline the overall component shape and add identification numbers to the component surface of a board. Figure 6-46 shows a photo of a board with component outlines and identification numbers. Component outlines and identification numbers make it easier to assemble the board and easier to identify components during board maintenance or troubleshooting.

Component outline drawings are prepared by tracing the master layout. They are done using tape and drawing film. The identifying numbers are added using dry-transfer numbers and letters, as shown in Figure 6-47. Figure 6-48 shows a component outline drawing. Note that the identification numbers are placed next to the appropriate component so that they can still be seen on the board after the component has been mounted.

## 6-10   INKING TECHNIQUES

Inked PC drawings are sometimes prepared because they are extremely durable and are well suited to photographic manufacturing techniques. Ink drawings are prepared using technical drawing pens such as the one shown in Figure 6-49.

Technical pens come in various widths labeled from 000 to 5, with 000 being the narrowest line. A pen is only able to produce one size of line, so a different line width will require a different pen. Drafters who do inking work usually carry several different-sized pens.

Two rules should be followed when preparing inked drawings: (1) hold the pen perfectly vertical, allowing the ink to flow out of the pen, and (2) always allow an air gap between the guiding instrument and the drawing medium. Unlike a ballpoint pen or a fountain pen, a technical pen requires very little pressure to operate. Once a line is started, the ink will continue to flow until the pen is lifted from the drawing. If a pen is simply stopped at the end of a line, a keyhole effect will result.

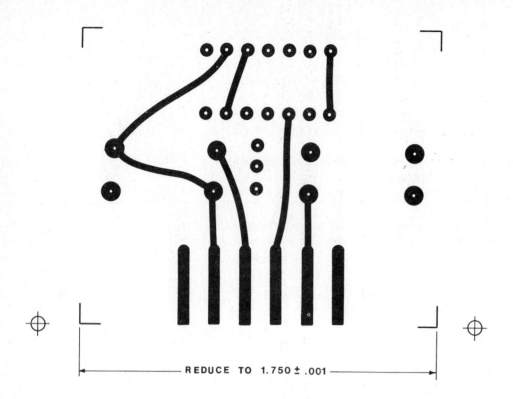

REDUCE TO 1.750 ± .001

COMPONENT SIDE

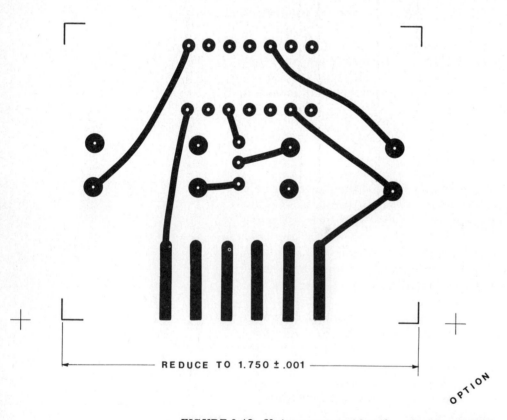

REDUCE TO 1.750 ± .001

OPTION

CONDUCTOR PATH SIDE

FIGURE 6-43   Various ways to identify a double-sided PC board.

146

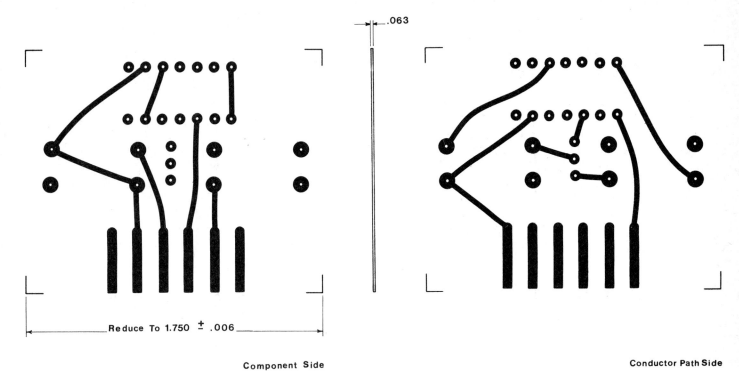

Component Side                                                                          Conductor Path Side

**FIGURE 6-44** Drawing that includes both sides of a double-sided PC board.

**FIGURE 6-45** (a) Pad master.

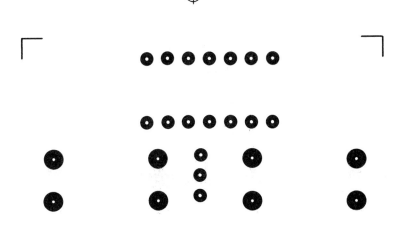

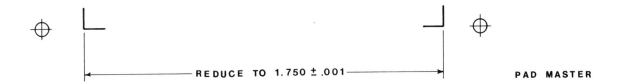

REDUCE TO 1.750 ± .001                                    PAD MASTER

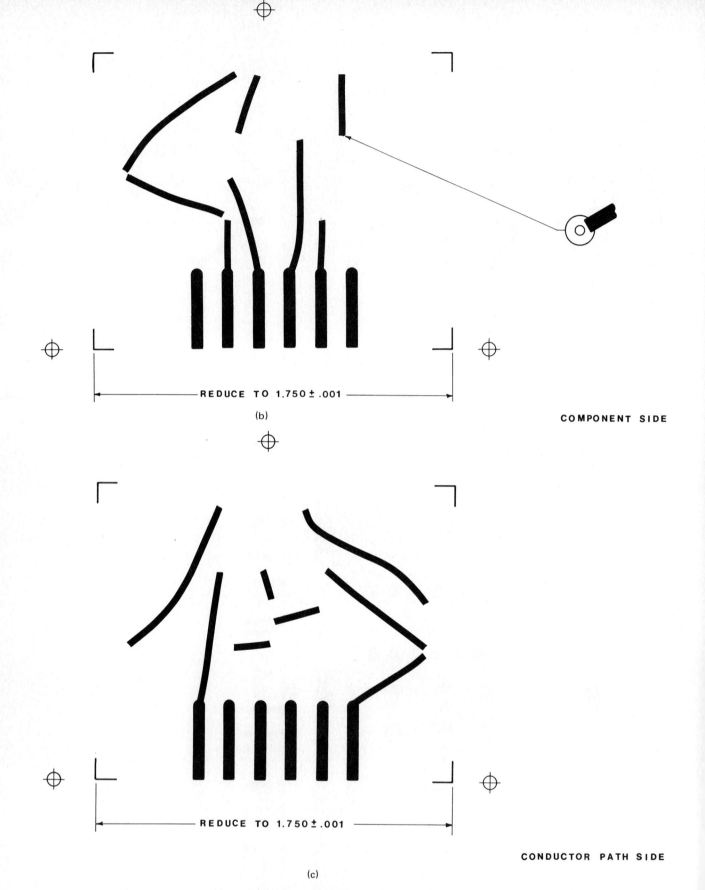

REDUCE TO 1.750 ± .001

(b)

COMPONENT SIDE

REDUCE TO 1.750 ± .001

(c)

CONDUCTOR PATH SIDE

FIGURE 6-45 (b) Component-side tape drawing; (c) conductor-path-side tape drawing.

**FIGURE 6-46** Component outlines and identification numbers.

**FIGURE 6-47** Applying dry-transfer letters to a drawing.

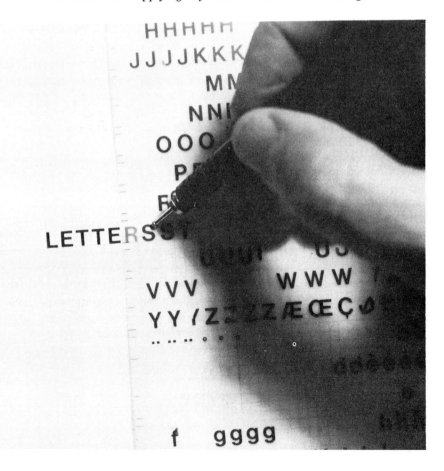

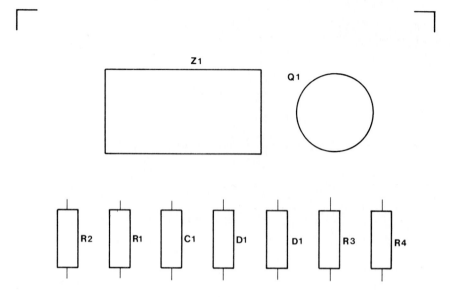

REDUCE **TO** 4.000 ± 0.001

**FIGURE 6-48** Component outline and identification number drawing.

**FIGURE 6-49** Technical drawing pens. (*Courtesy of B.L. Makepeace Co., Boston.*)

61 0051-00

61 0053-0
Jewel Tip
Reservoir Pen
in 61 0055
Holder

61 0062

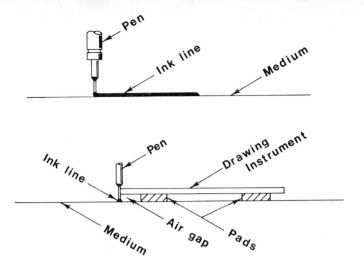

FIGURE 6-50 Drawing instruments must be off the drawing surface when inking.

An air gap (see Figure 6-50) is required to keep the ink from flowing under guiding instruments. An air gap is created by lifting the instrument off the drawing surface by sliding a triangle or template under the instrument to be used for the drawing. It may also be done by attaching pressure pads (circular pieces of plastic with glue on one side) or by taping pennies to the drawing instrument.

Figure 2-11 pictures a Leroy inking setup. Leroy is a tracing technique that uses a bar with symbols engraved and a scriber. The scriber has two points and an inkwell. One point rides in a groove at the bottom of the bar and references the scriber. The second point is used to trace the engraved symbols, and the inkwell in turn follows the second point, copying the traced symbol onto the drawing. A technical pen can be used in place of the inkwell.

## PROBLEMS

6-1 through 6-9   Based on the schematic diagrams in Figures P6-1 through P6-9, prepare the following types of drawings.

   a. A single-sided space allocation drawing
   b. A double-sided space allocation drawing
   c. A tape master
   d. Soldering masks
   e. A component outline drawing
Use the component sizes as defined in Appendix H.

**FIGURE P6-1**

MIC PREAMPLIFIER

151

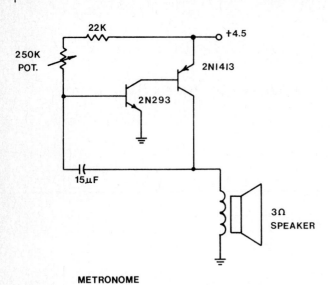

METRONOME

**FIGURE P6-2**

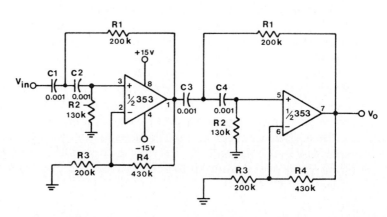

**FIGURE P6-3**

**FIGURE P6-4**

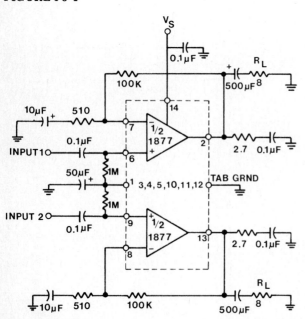

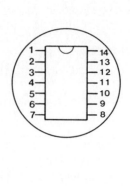

FIGURE P6-5

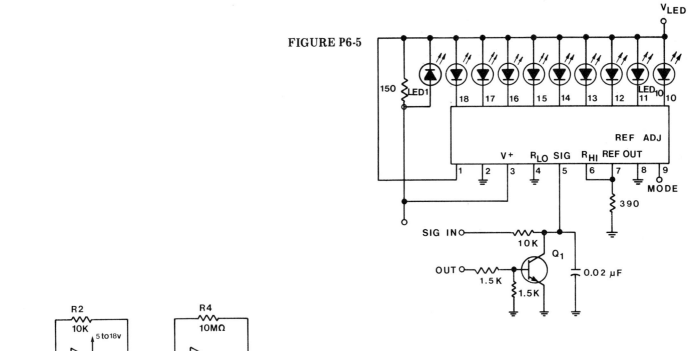

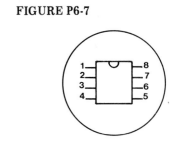

FIGURE P6-7

555 TIMER

FIGURE P6-6

C1 stores the peak voltage at $V_{in}$

1458 DUAL OPERATIONAL
AMPLIFIER

Warble Alarm Circuit

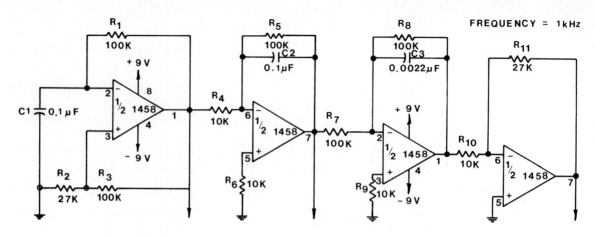

FREQUENCY = 1 kHz

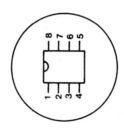

1458 DUAL OPERATIONAL
AMPLIFIER

**FIGURE P6-8**

**FIGURE P6-9**

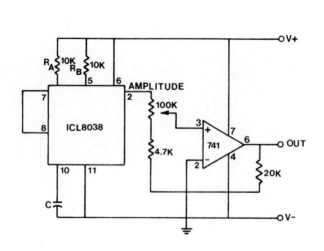

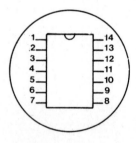

ICL8038 PRECISION
WAVEFORM GENERATOR

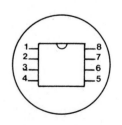

741 OPERATIONAL AMP

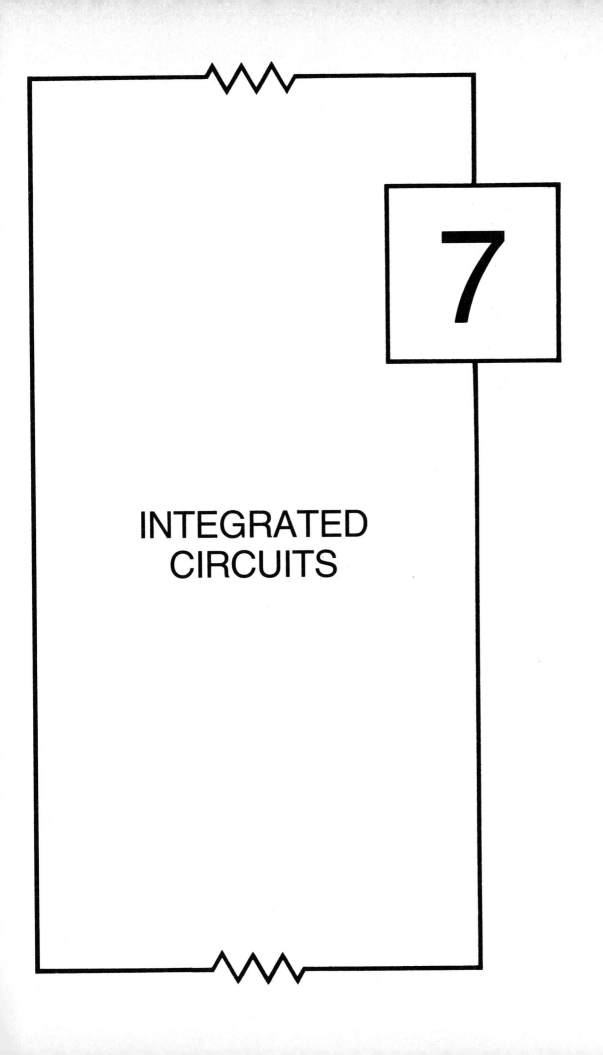

# 7

# INTEGRATED CIRCUITS

## 7-1   INTRODUCTION

In this chapter we explain how integrated circuits (ICs) are manufactured and what drafting procedures are required in the manufacturing process. Some IC design concepts are presented, but only in the depth needed to help understand how to create appropriate masks.

There are several terms which are used when speaking about ICs.

*Bit:*  Binary digit, the smallest part of an IC. It has either high voltage or low voltage, expressed as either 1 or 0.

*Byte:*  8 bits.

*EPROM:*  Erasable programmable read-only memory.

*Gate:*  The controlling element of certain transistors, or a basic logic circuit.

*K:*  1024 bits (the value comes from the binary system). A 16K system has 16 × 1024, or 16,384 bits of information.

*Logic:*  The order in which an IC is designed so as to produce certain outputs from given inputs.

*LSI:*  Large-scale integration.

*Mask:*  Plates used during manufacturing to define the required logic pattern.

*Memory chip:*  An IC that stores information using electrical charges.

*Microprocessor:*  An IC that performs many different functions.

*RAM:*  Random access memory; memory that can be changed by an operator.

156

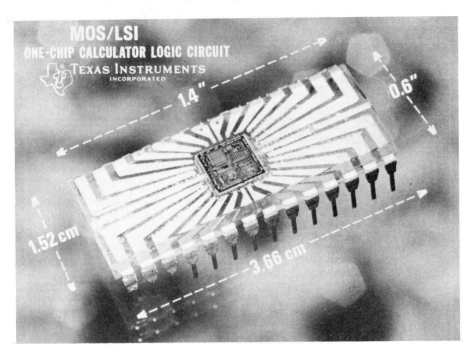

**FIGURE 7-1**   Integrated circuit. (*Courtesy of Texas Instruments, Inc.*)

*ROM:*   Read-only memory; memory that can not be changed (e.g., a digital calendar).

*Transistor:*   Functions as either an amplifier or a current switch in an IC.

*VLSI:*   Very large scale integration.

*Wafer:*   The thin disk on which an IC is manufactured.

Figure 7-1 shows an integrated circuit.

## 7-2   HOW AN IC IS MANUFACTURED

An IC is manufactured on a thin piece of silicon called a wafer. The manufacturing of silicon wafers is called "growing" the purified silicon. The wafers are then cut into the desired shape and coated, first with an oxide, then with a coating of photoresist (see Figure 7-2).

Photoresist material is light sensitive. If light is shown on photoresist material, the material becomes hard and will resist the acids and solvents used in etching. Conversely, the photoresist material *not* exposed to light (covered by a mask) will remain soft and can be removed by etching.

The treated silicon wafer is then exposed to light as shown in Figure 7-3. Areas exposed to light will harden. Areas blocked from the light by the mask will remain soft. The coated wafer is then etched, removing the soft photoresist.

The oxide layer is now processed by hot gases. The gas will not penetrate the hardened photoresist, but will enter the open areas created during etching, resulting in cavities in the oxide layer as shown.

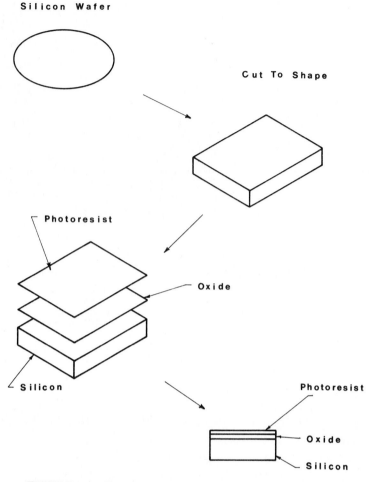

**FIGURE 7-2** How a silicon wafer is prepared for IC manufacture.

The process is then repeated by first removing the hardened photoresist material and applying a new layer of silicon. A new layer of photoresist and oxide is attached, then etched as needed. The result will be a gradual buildup of cavities and filled areas which serve to form the IC chip.

The conductor zones are interconnected by again coating the wafer with silicon and photoresist. The open areas that penetrate the silicon layers to the conductor zones are called *windows* and are created using a masking technique. Metal is then condensed into the windows to form the required connections between positive and negative zones.

The entire process is dependent on correctly prepared masks. Drafters are usually responsible for mask preparation.


## 7-3   CHOOSING A SCALE

Masks for IC manufacturing are done at scales from 200 to 1000:1, with 400 or 500:1 being most common. Very large scales are required because the final sizes are so small.

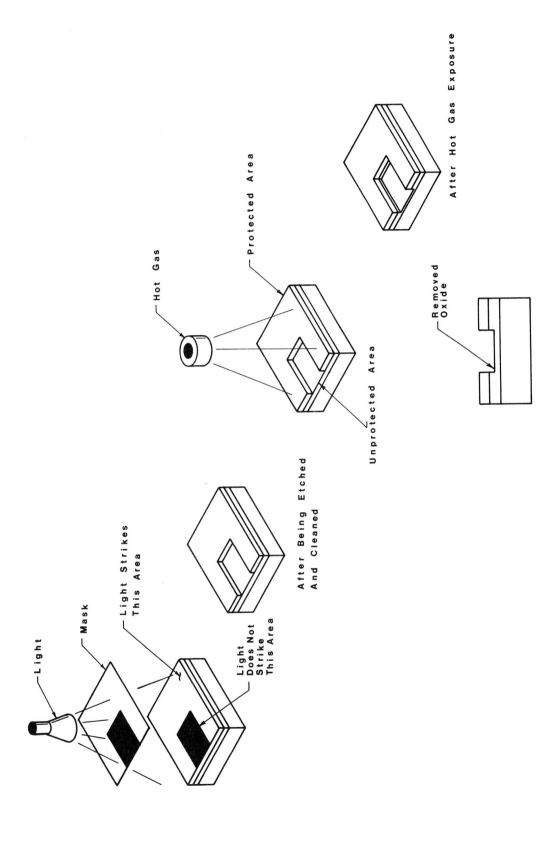

Light

Mask

Light Strikes
This Area

Light
Does Not
Strike
This Area

After Being Etched
And Cleaned

Hot Gas

Protected Area

Unprotected Area

After Hot Gas Exposure

Removed
Oxide

FIGURE 7-3  How an IC is manufactured.

159

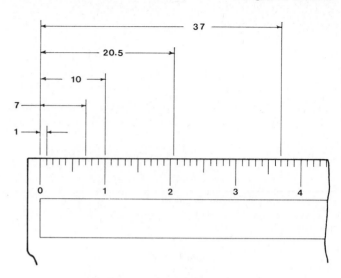

FIGURE 7-4   Sample metric scale.

$1\ \mu m = 1 \times 10^{-6}\ m$
$1\ mm = 1 \times 10^{-3}\ m$
$25.4\ mm = 1.00\ in$
$0.001\ in = 2.54\ \mu m$

$\mu$ = micro
m = meter
mm = millimeter
in = inch

FIGURE 7-5   Equivalent values.

Measurements for IC masks are usually done using millimeters (mm). Figure 7-4 shows a metric scale together with some sample measurements.

A millimeter is one thousandth of a meter. There are 10 millimeters per centimeter and 100 centimeters in a meter.

IC mask layouts use sizes measured in micrometers ($\mu$m). One thousandth of an inch equals twenty-five and four-tenths micrometers (0.001 inch = 25.4 $\mu$m).

One thousandth of an inch is called a mil (0.001 inch = 1 mil). Be careful to understand which units are being specified. Figure 7-5 shows some equivalent values.

Measurements are converted to 400 or 500 scale values by multiplying the measurement by the appropriate scale value. For example, 5 $\mu$m at 400 scale would be 5(400) = 2000 $\mu$m. This means that the mask value would be 2 mm (2000 × 0.001 = 2). Figure 7-6 gives some additional examples.

FIGURE 7-6   How to convert a given value to a larger scale.

( value ) ( scale ) ( .001 ) = mask

$( \ 5\ \mu m ) ( \ 400 ) ( .001 ) = 2.0\ mm$
$(15\ \mu m ) ( \ 500 ) ( .001 ) = 7.5\ mm$
$( \ 8\ \mu m ) (1000 ) ( .001 ) = 8.0\ mm$

## 7-4   TRANSISTORS

Transistors are developed in IC circuits as shown in Figure 7-7. Figure 7-7 shows the symbol for an NPN transistor. The three terminals c, b, and e are also labeled with their IC equivalents: $N^+$, P, and $N^+$. Step 1 shows the development of the $N^+$ and P regions. The dimensions for the regions will vary with speed requirements and voltage intensity for the circuit.

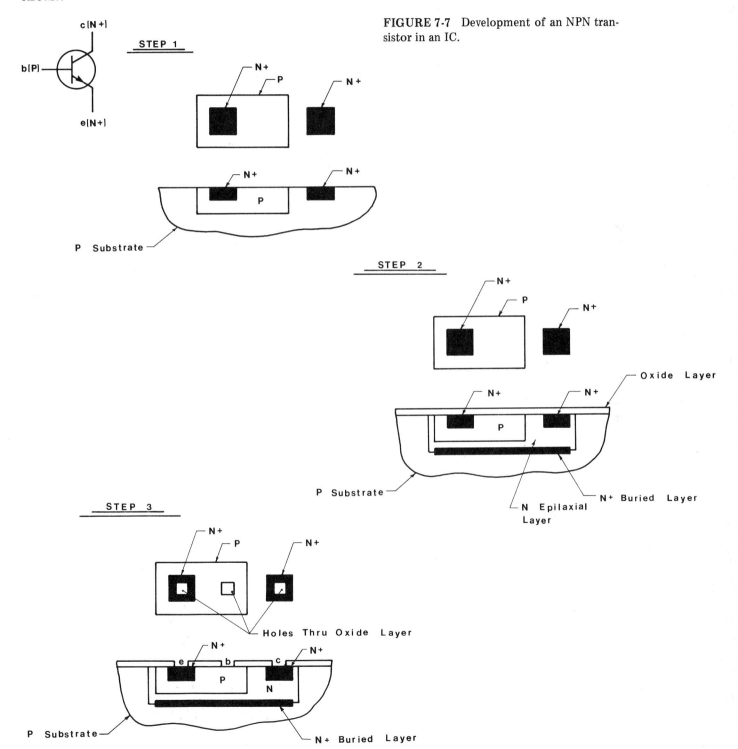

**FIGURE 7-7**  Development of an NPN transistor in an IC.

Step 2 shows the addition of an $N^+$ buried layer which helps reduce resistance between the N epilaxial layer and the P substrate. The N epilaxial layer is added after the $N^+$ buried layer. A new oxide layer is also added.

In step 3, holes are added through the oxide layer. These holes are equivalent to the c, b, and e terminals and are used to connect the transistor to the rest of the circuit.

Figure 7-8 shows how masks are used to create a transistor. Figure 7-8(a) shows an isolation area mask. This mask isolates the transistor and prevents electrical interference with other parts of the circuit.

Figure 7-8(b) shows the mask used to create the $N^+$ buried layer. The N epilaxial layer does not require a separate mask. It is added on top of the buried layer.

**FIGURE 7-8** (a) Isolation area mask; (b) $N^+$ buried-layer mask; (c) P base layer mask; (d) $N^+$ layer mask; (e) contact hole mask; (f) metalization layer mask.

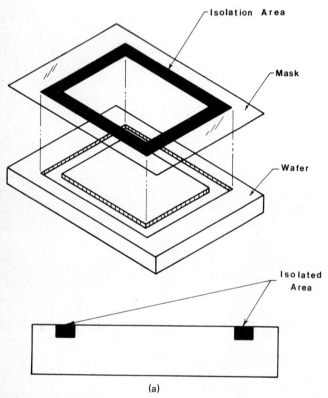

(a)

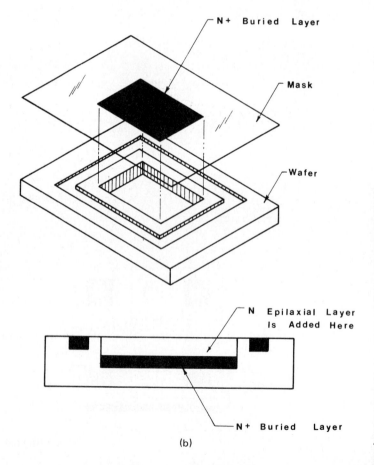

(b)

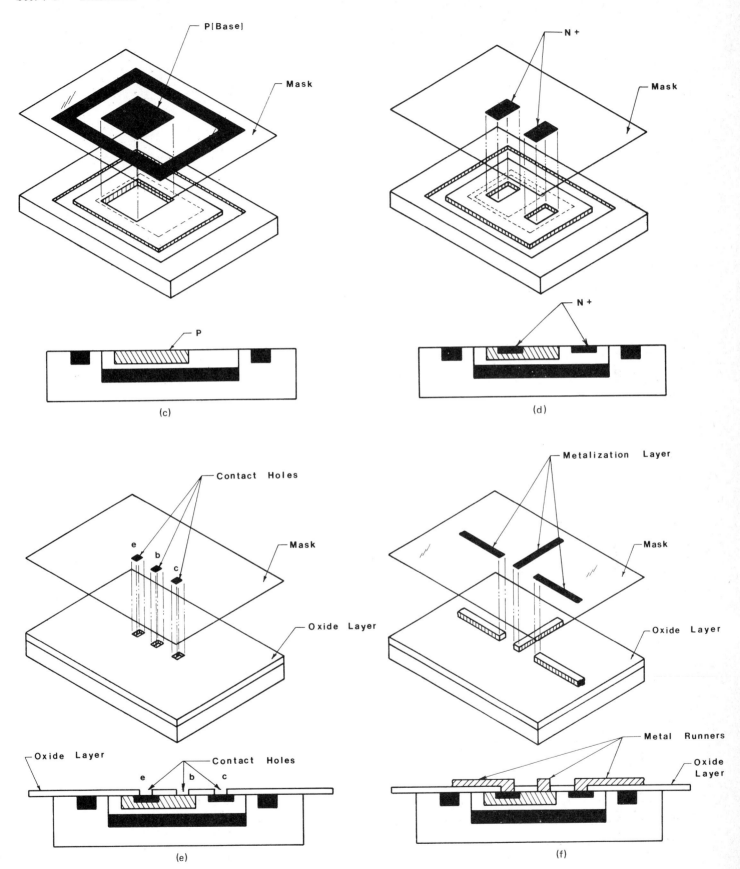

**FIGURE 7-8**  (cont.)

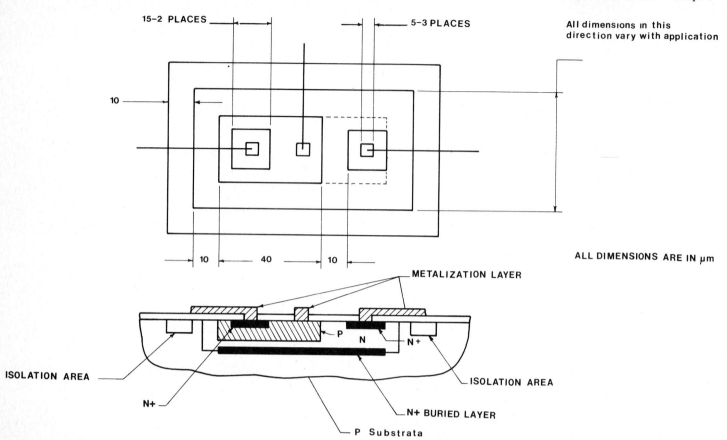

**15-2 PLACES**         **5-3 PLACES**

10

**ALL DIMENSIONS ARE IN μm**

10     40     10

**METALIZATION LAYER**

P    N     N⁺

**ISOLATION AREA**

**ISOLATION AREA**

N⁺

**N⁺ BURIED LAYER**

**P Substrata**

**FIGURE 7-9** Master layout for an IC transistor. The dimensions presented represent one of many different layout sizes.

Figure 7-8(c) shows the addition of the P base layer. The isolation area is repeated in this mask to help reinforce the isolation area.

Figure 7-8(d) shows the mask for the $N^+$ areas, and Figure 7-8(e) shows the mask for the windows or contact hole.

Figure 7-8(f) shows the metalization mask. This mask is used to connect the transistor to the rest of the circuit. Metal, usually aluminum, is added in this step. The runners are broken in the mask, as they would continue into other areas of the circuit.

Masks are created from a master layout as shown in Figure 7-9. The master layout is used to space the various layers. The hidden lines represent the buried layer. The dimensions given are only representative of a typical transistor layout. Dimensions will vary with application. Figure 7-10 shows a transistor master layout dimensioned in mils.

Transistors may be positioned next to each other as shown in Figure 7-11. Note in the equivalent circuit that the e of Q1 is connected to the c of Q2. This connection is shown as a step in the master layout. This is because most layouts are done on computer graphics systems which can draw only horizontal and vertical lines. Note that the two transistors are separated by an isolation area.

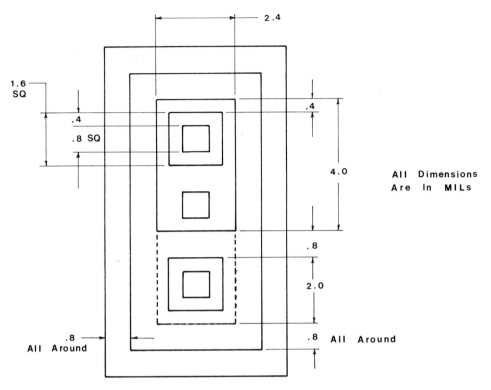

**FIGURE 7-10**  Typical transistor master dimensioned in mils.

**FIGURE 7-11**  Two N-P-N transistors.

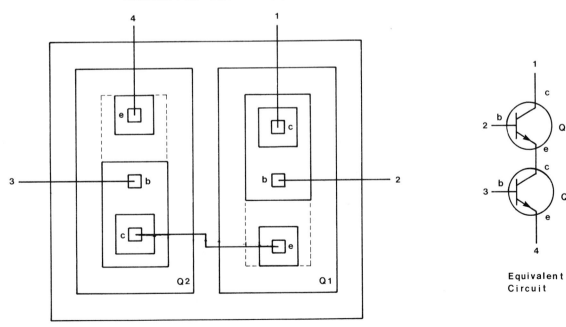

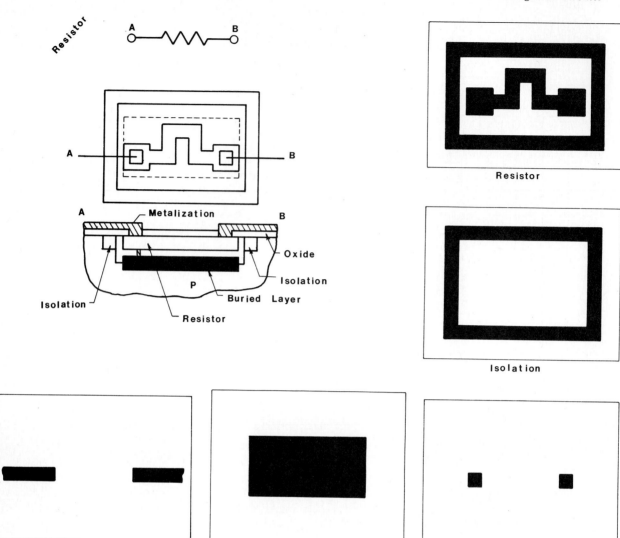

**FIGURE 7-12**  Resistor in an IC together with the appropriate masks.

## 7-5   RESISTORS

Resistance in integrated circuits can be created using the base material, the emitter material, the collector material, or by a process called pinching. This section deals with base material resistance, which is the most widely used process. Figure 7-12 shows a base diffused resistor.

The masks needed to create the resistor are also shown. Note that an N$^+$ buried layer is also included and that the resistor is isolated. Further, the isolation is repeated in the buried layer mask to help to reinforce it.

The amount of resistance depends on the resistance of the material used, the material's cross-sectional area, and the material's length. The

·resistance of the material is called *sheet resistance* and is measured in *ohms per square* ($\Omega/\square$). The total value of a resistor is calculated using the formula

$$R = R_s \left(\frac{l}{w}\right) + \text{corners} + \text{contacts}$$

where $R_s$ = sheet resistance
$l$ = length of sheet
$w$ = width of sheet

Figure 7-13 shows these terms. For simplicity, corners and contacts generally are valued at one-half a sheet square. So, for example, if the sheet resistance is 200 $\Omega/\square$, the corner and the contact area would each be worth 100 $\Omega/\square$.

Figure 7-14 shows a resistance between points A and B. The value of the resistance is calculated as follows:

$$R = R_s \left(\frac{l}{w}\right) + \text{corners} + \text{contacts}$$

Given $R_s$ = 200 $\Omega/\square$, we have

$$R = (200) \frac{10 + 10 + 10 + 25 + 10 + 25 + 10 + 10 + 10}{5}$$

$$+ \text{8 corners} + \text{2 contacts}$$

$$= 4800 + 800 + 200$$

$$= 5800 \text{ k}\Omega$$

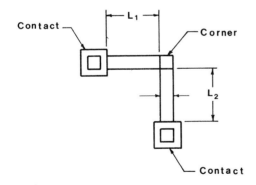

**FIGURE 7-13**  Resistor terms.

**FIGURE 7-14**  Resistance between points A and B.

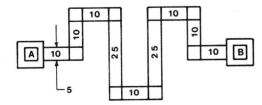

SCHEMATIC  DIAGRAM

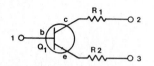

MASTER  LAYOUT

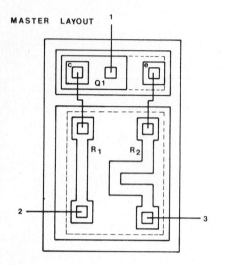

MASKS

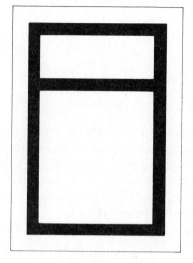

Isolation

**FIGURE 7-15** IC that has two transistors and a resistor, together with the appropriate masks.

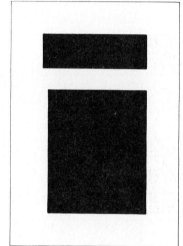

Buried   Layer

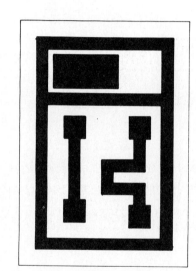

BASE

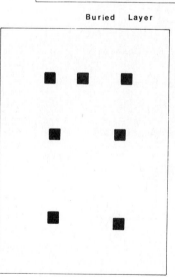

EMITTER

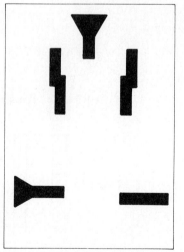

CONTACT

METALIZATION

Figure 7-15 shows a circuit that has a transistor and two resistors. The master layout and masks have also been included. The transistor is isolated, but both resistors are located within the same region. The metal runners cross between the isolation area connecting the various terminal windows.

## 7-6   DIODES AND CAPACITORS

Figure 7-16 shows an IC diode both as a symbol and as it would appear on an IC chip. The various layers are created using the masking techniques shown for transistors and resistors. Diodes with common anodes or cathodes may be combined to save space. Figure 7-16 shows an example of a master layout for two diodes with a common cathode.

Figure 7-17 shows an approximate representation of a master layout for a capacitor. There is a continuous relationship between capacitors and diodes in IC chips called a parasitic relationship. The actual design and subsequent layout of capacitors is beyond the scope of this book.

**FIGURE 7-16**   Diodes in an IC.

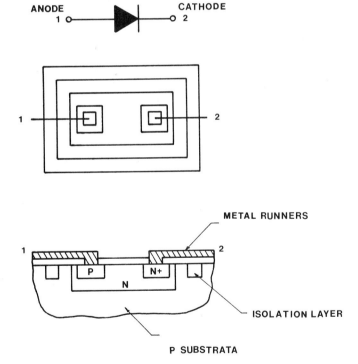

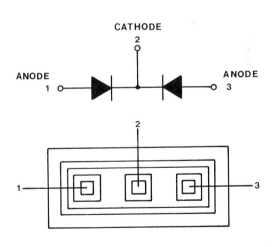

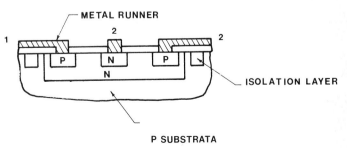

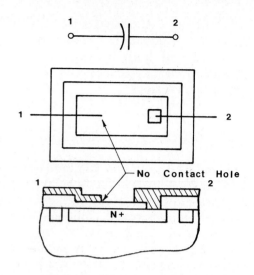

**FIGURE 7-17** Approximate capacitor in an IC.

THIS IS ONLY AN APPROXIMATE REPRESENTATION

## PROBLEMS

**7-1** Convert each value from μm to mils.
   a. 10
   b. 50
   c. 5
   d. 15
   e. 20

**7-2** Convert each value from mils to μm.
   a. 1.0
   b. 2.0
   c. 1.7
   d. 1.5
   e. 2.5

**7-3** Redraw and complete the table shown in Figure P7-3. Include units for each value.

**FIGURE P7-3**

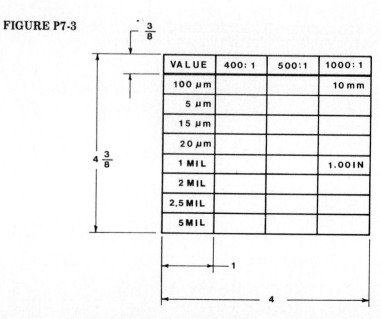

| VALUE | 400:1 | 500:1 | 1000:1 |
|---|---|---|---|
| 100 μm | | | 10 mm |
| 5 μm | | | |
| 15 μm | | | |
| 20 μm | | | |
| 1 MIL | | | 1.00 IN |
| 2 MIL | | | |
| 2.5 MIL | | | |
| 5 MIL | | | |

1 MIL = 25.4 μm

7-4    Redraw Figure P7-4 at a scale of:
a.    400:1
b.    500:1
c.    1000:1

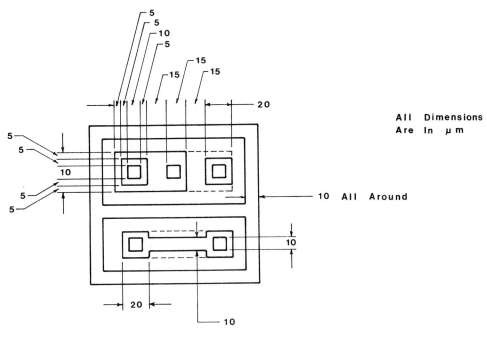

All    Dimensions
Are    In    μm

**FIGURE P7-4**

7-5    Calculate the resistance for the shape shown in Figure P7-5. Assume that the material has a sheet resistance of 200 Ω/□.

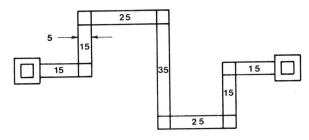

**FIGURE P7-5**

7-6    Calculate the resistance for the shape shown in Figure P7-6. Assume that the material has a sheet resistance of 200 Ω/□.

**FIGURE P7-6**

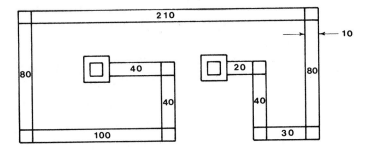

**7-7**   Design a resistance pattern that generates a resistance of 20 k$\Omega$ $\pm$ 10%. Assume that the two end points, A and B, are located 100 $\mu$m from each other.

**7-8**   Use the master layout shown in Figure P7-8 to prepare a complete set of six masks. Make the metal runners equal in width to the contact windows. Assume that the layout is drawn to scale.

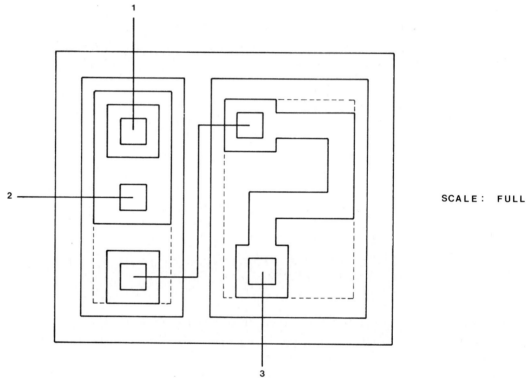

SCALE:  FULL

**FIGURE P7-8**

**7-9**   Use the master layout shown in Figure P7-9 to prepare a complete set of six masks. Make the metal runners equal in width to the contrast windows. Assume that the layout is drawn to scale.

**7-10**   Make a complete set of five masks from the master layout of the resistor shown in Figure P7-10. Make the metal runners equal in width to the contact windows. Assume that the layout is drawn to scale.

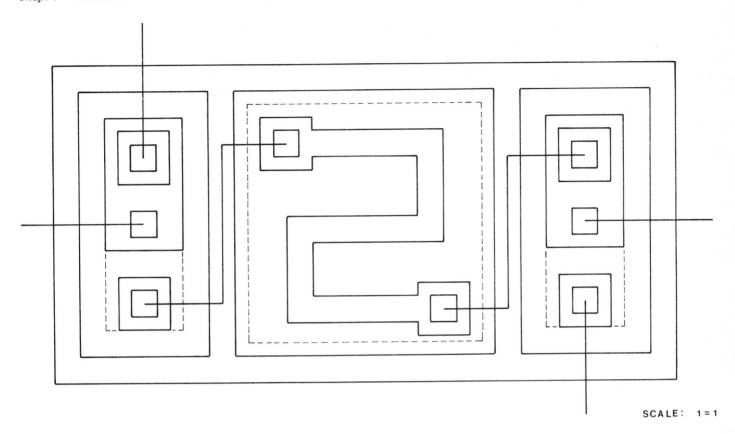

FIGURE P7-9

FIGURE P7-10

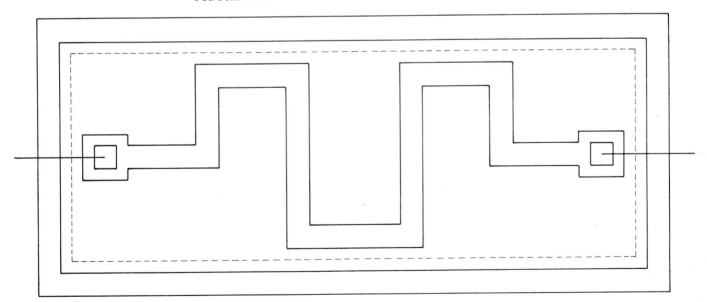

SCALE:   1=1

**7-11**  Prepare a master layout and appropriate masks (a set of six) for the schematic diagram shown in Figure P7-11. Use the general dimensions shown in Figure P7-10 to size the transistors. Assume that the resistance material is 10 $\mu$m wide and has a sheet value of 200 $\Omega/\square$. Assume that the end area is 15 $\mu$m square.

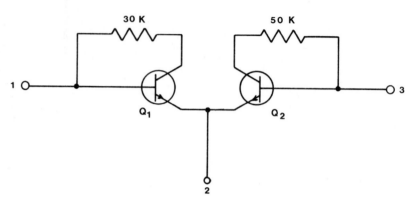

**FIGURE P7-11**

**7-12**  Repeat Problem 7-11 for the schematic shown in Figure P7-12.

**FIGURE P7-12**

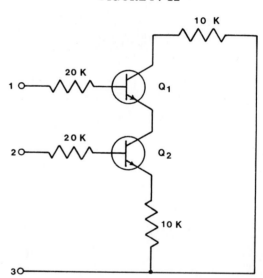

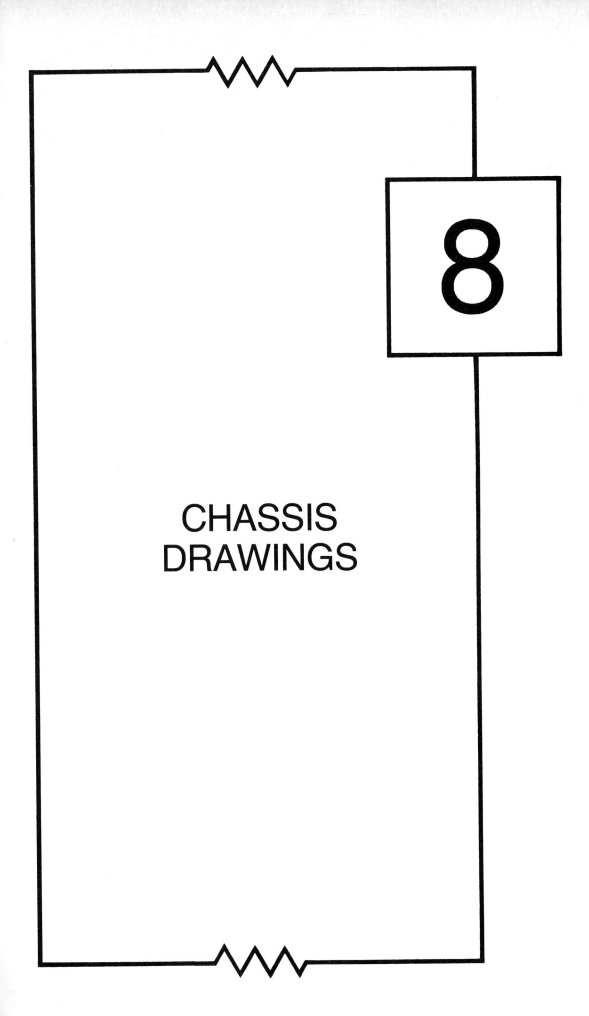

# 8

# CHASSIS
# DRAWINGS

## 8-1 INTRODUCTION

In this chapter we study the design and drawing of chassis. Included are an explanation of the various drawing techniques used to draw chassis, an explanation of how chassis flat patterns are developed, and a review of some of the fasteners used in chassis design.

## 8-2 TYPES OF CHASSIS

There are many different materials and shapes which are used to create a wide variety of chassis designs. General categories of chassis designs are usually identified by the basic shape. For example: box, U, T, and I chassis are designs which are shaped like a box, U, T, or I, respectively. Figure 8-1 shows a U-type chassis with flanges. Chassis may be purchased from commercial manufacturers or they may be fabricated in a local shop.

## 8-3 FLAT PATTERN DESIGN

Chassis are fabricated by first cutting a flat pattern from the desired material and then bending the flat pattern into the final chassis shape. As metal is bent, it stretches, which means that the length of the flat pattern must be less than the finished overall length of the chassis.

The correct size of a flat pattern is determined by considering the overall size requirements and the amount of material for *bend allowance*. Bend allowance is the amount of material needed to form bends

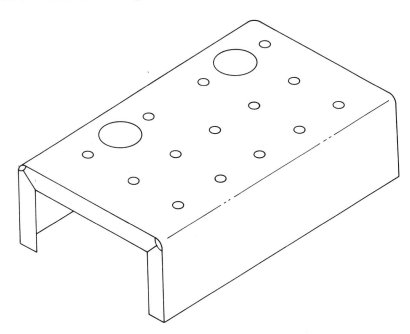

**FIGURE 8-1**   U-type chassis with flanges.

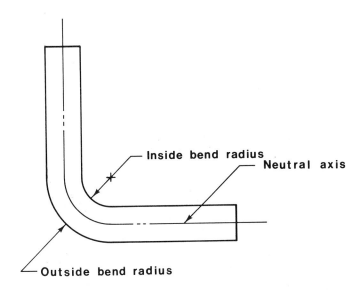

**FIGURE 8-2**   Illustration of inside bend radius, neutral axis, and outside bend radius.

and includes a consideration of how the material will stretch during the bending. Bend allowances and the lengths of straight sections are added together to determine the diminished length of the flat pattern.

Before proceeding with a discussion of how to determine flat pattern lengths, we must first understand three terms: *inside bend radius*, *outside bend radius*, and *neutral axis*. Figure 8-2 defines the terms. The neutral axis is a theoretical line exactly halfway between the inside bend radius and the outside bend radius. Material between the neutral axis and the inside bend radius is under compression during bending and material between the neutral axis and the outside bend radius is under tension (being stretched).

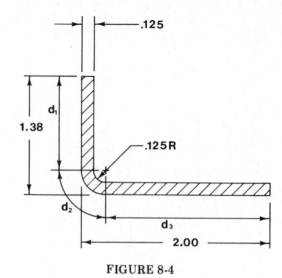

FIGURE 8-3

FIGURE 8-4

The difference between the inside and outside bend radii is equal to the material thickness:

$$\text{inside bend radius} + \text{material thickness}$$
$$= \text{outside bend radius} \tag{F-1}$$

Drawings usually define only inside bend radius and material thickness, so if the value of the outside bend radius is desired, it must be calculated using Formula (F-1). In Figure 8-3 the inside bend radius is $\frac{3}{16}$ inch and the material thickness is $\frac{1}{8}$ inch. Applying these data to Formula (F-1), we get

$$\frac{3}{16} + \frac{1}{8} = \text{outside bend radius}$$

$$\frac{3}{16} + \frac{2}{16} = \text{outside bend radius}$$

$$\frac{5}{16} = \text{outside bend radius}$$

The length of a flat pattern is determined by adding the length of the straight sections of the final chassis shape to the lengths of the curved sections, which have been modified to account for bend allowance. If we apply this concept to the contour shown in Figure 8-4, we see that

$$\text{flat pattern length} = d_1 + d_2 + d_3$$

The distance $d_1$ is found by taking the overall length and subtracting the inside bend radius and material thickness:

$$d_1 = 1.38 - 0.125 - 0.125$$
$$= 1.13$$

Similarly,

$$d_3 = 2.00 - 0.125 - 0.125$$
$$= 1.75$$

The distance $d_2$ must be calculated considering the bend allowance. This is done using the formula*

$$B = \frac{A}{360} \, 2\pi \, (\text{IR} + Kt) \qquad\qquad (\text{F-2})$$

where $B$ = bend allowance

$\qquad A$ = bend angle

$\qquad$ IR = inside bend radius

$\qquad K$ = constant

$\qquad t$ = material thickness

For the problem presented in Figure 8-4,

$$A = 90°$$

$$\text{IR} = 0.125$$

$$K = 0.33$$

$$t = 0.125$$

The value of $K$ is equal to 0.33 when the inside bend radius is less than $2t$ (two times the material thickness) and is equal to 0.50 when the inside bend radius is greater than $2t$. In our problem,

$$2t = 2(0.125) = 0.25$$

which means that

$$K = 0.33$$

as the inside bend radius is 0.125, which is less than 0.25. Subtracting the values in Formula (F-2), we have

$$B = \frac{90}{360} \, (2\pi)[0.125 + (0.33)(0.125)]$$

$$= 1.57(0.125 + 0.041)$$

$$= 0.26$$

Therefore,

$$d_2 = 0.26$$

We can now calculate the flat pattern length by adding $d_1, d_2,$ and $d_3$:

$$\text{flat pattern length} = d_1 + d_2 + d_3$$

$$= 1.13 + 0.26 + 1.75$$

$$= 3.14$$

Figure 8-5 is another example of how to calculate the flat pattern length for a given chassis shape.

---

*From ASTM *Die Design Handbook*.

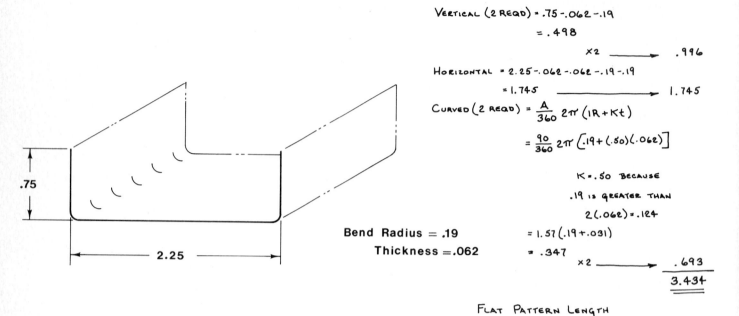

STRAIGHT SECTIONS

VERTICAL (2 REQD) = .75 - .062 - .19

$\qquad$ = .498

$\times 2 \longrightarrow$ .996

HORIZONTAL = 2.25 - .062 - .062 - .19 - .19

$\qquad$ = 1.745 $\longrightarrow$ 1.745

CURVED (2 REQD) = $\frac{A}{360} 2\pi (IR + Kt)$

$\qquad = \frac{90}{360} 2\pi \left[ .19 + (.50)(.062) \right]$

K = .50 BECAUSE

.19 is GREATER THAN

2(.062) = .124

Bend Radius = .19    $= 1.57 (.19 + .031)$

Thickness = .062    $= .347$

$\times 2 \longrightarrow$ .693

3.434

.75

2.25

FLAT PATTERN LENGTH

= 3.434

**FIGURE 8-5**  Example of a flat-pattern length calculation.

## 8-4   HOW TO DRAW CHASSIS

Chassis are sometimes difficult to draw because they contain few square corners and are made from thin materials, which makes drawing hidden lines confusing. These difficulties may be overcome by applying the following drawing techniques.

Round corners are easier to draw if we first draw the object as shown in Figure 8-6. Then darkening in the object lines, start by first darkening in the round corners. This will make it easier to assure an even, continuous line both into and out of the corner. Also, remember when drawing round corners that the labels on a circle template are diameters, whereas bending values are defined in radii. Do not forget to multiple radius values by 2 before using a circle template.

2(radius) = diameter

Hidden lines are drawn in sheet metal parts as shown in Figure 8-7. In every case, the centerline for a hole is included. If the material is so thin that the convention pictured in Figure 8-7 is not practical, an enlarged detail should be considered.

**FIGURE 8-6**  How to draw shapes made from thin materials.

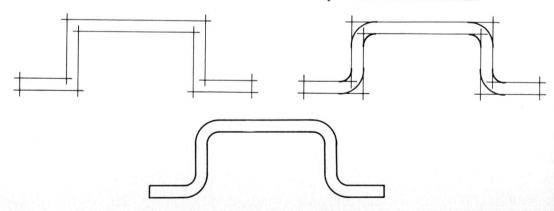

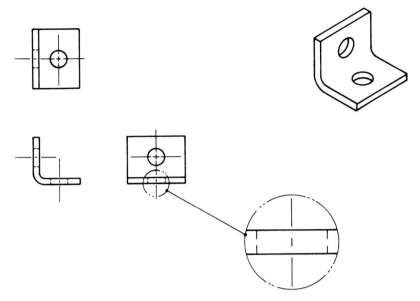

FIGURE 8-7   Holes in thin materials.

## 8-5   HOW TO DIMENSION CHASSIS DRAWINGS

Chassis designs are characterized by many small holes in a small area. Conventional dimensioning techniques would tend to create drawings that would be a maze of numbers, extension, dimension, and leader lines. To help produce neat, clear, and easy-to-read chassis drawings, two different dimensioning systems are commonly used: the baseline system and the coordinate system.

The *baseline system* refers all dimensions to common baselines. These lines are sometimes called datum lines or reference lines. The baseline system is illustrated in Figure 8-8. The baseline system has the advantage of eliminating cumulative tolerance errors. Each dimension is taken independently and an error in one dimension will not carry over to other dimensions. The disadvantages of the baseline system are that

FIGURE 8-8   Example of baseline system dimensioning.

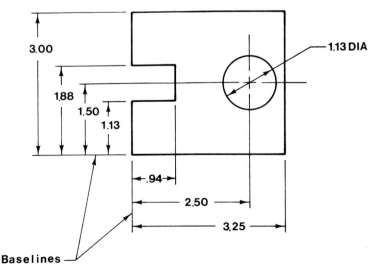

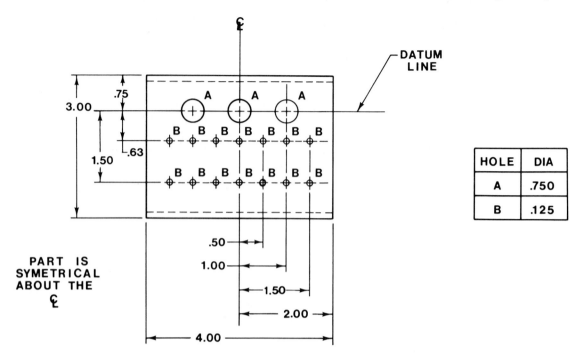

FIGURE 8-9   Example of baseline system dimensioning.

it requires a large amount of area to complete, usually at least twice the area of the object, and that once the dimensions are in place on the drawing, it is very difficult to make changes or additions.

Most chassis have rounded edges which do not make good dimensioning reference lines because they are difficult to align consistently. Most drafters use a centerline or some other line within the object as a reference line. Note the location of the baselines in Figure 8-9.

To use the baseline system (see Figure 8-10):

1.  Prepare a scale drawing of the chassis surface locating all holes. Be sure to include centerlines for every hole.
2.  Define the baselines. Each surface will require two baselines.
3.  Locate all holes and edges from the baselines. Space the dimension lines ¼ inch apart and place dimensions in consecutive order according to length starting with the shortest dimension closest to the baseline.
4.  Define the hole sizes by assigning a letter to each hole size and then defining each letter in a chart, as shown in Figure 8-10. Place the letters on the drawing to the right and above the appropriate hole, if possible. Keep the letters close to the holes they define.

The *coordinate system* is based on a 90° *x-y* coordinate system and is particularly well suited for use when programming numerically controlled machines. The coordinate system is easier to draw than the baseline system and is easier to change or correct but does require the reader to look up each dimension. Figure 8-11 illustrates a chassis surface dimensioned using the coordinate system.

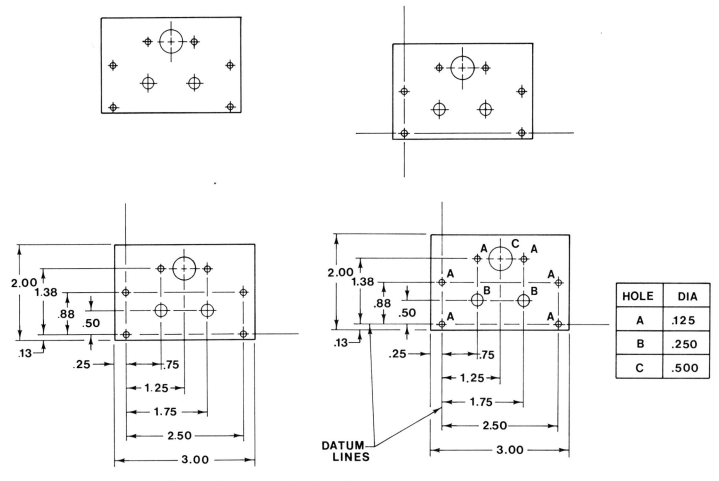

**FIGURE 8-10** How to dimension a chassis surface using the baseline system.

| HOLE | DIA |
|------|------|
| A | .125 |
| B | .250 |
| C | .500 |

**FIGURE 8-11** Example of coordinate system dimensioning.

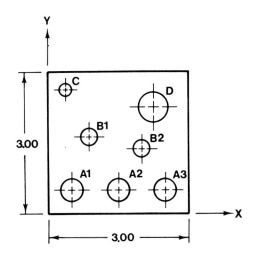

| HOLE | X | Y | DIA |
|------|------|------|------|
| A1 | 0.50 | 0.50 | .470 |
| A2 | 1.50 | 0.50 | .470 |
| A3 | 2.50 | 0.50 | .470 |
| B1 | 0.88 | 1.62 | .375 |
| B2 | 2.00 | 1.38 | .375 |
| C | 0.38 | 2.62 | .250 |
| D | 2.25 | 2.25 | .625 |

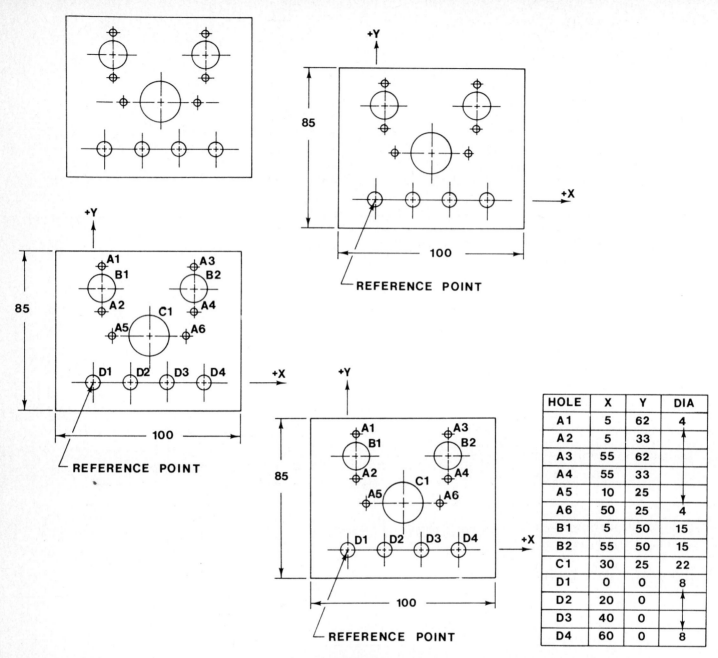

| HOLE | X | Y | DIA |
|------|----|----|-----|
| A1 | 5 | 62 | 4 |
| A2 | 5 | 33 | |
| A3 | 55 | 62 | |
| A4 | 55 | 33 | |
| A5 | 10 | 25 | |
| A6 | 50 | 25 | 4 |
| B1 | 5 | 50 | 15 |
| B2 | 55 | 50 | 15 |
| C1 | 30 | 25 | 22 |
| D1 | 0 | 0 | 8 |
| D2 | 20 | 0 | |
| D3 | 40 | 0 | |
| D4 | 60 | 0 | 8 |

FIGURE 8-12 How to dimension using the coordinate system.

To use the coordinate system (see Figure 8-12):

1. Prepare a scale drawing of the chassis surface locating all holes. Be sure to include centerlines for every hole.

2. Define the reference point and then clearly label the positive $x$ and $y$ axes. Add the overall dimensions.

3. Label each hole using a letter and a number. Assign the same letter, but different consecutive numbers, to holes of equal diameter. In the example problem, the holes labeled $A$ are all 4 mm in diameter; holes labeled $B$ are 10 mm in diameter; and so on.

4. Prepare a chart which includes a listing, using the labels assigned in step 3, of the distance that each hole is from the $x$ and $y$ axes and the hole's diameter.

## 8-6   CHASSIS FASTENERS

There are several different types of fasteners used to manufacture chassis, including rivets, sheet metal screws, and machine screws. Welded and soldered joins are also used. Thread representations were discussed in Section 1-10 and these representations are also valid for threaded chassis fasteners.

Rivets are shown on a drawing by using the representations shown in Figure 8-13. There are two types of representations: detailed and schematic. Rivets can be identified by using the coding system of the National Aircraft Standards (NAS), which is illustrated in Figure 8-14.

Provided that the rivets are all exactly the same, a long row of rivets may be called out by calling out only the first and last rivets in the row. Figure 8-15 illustrates this kind of rivet callout.

**Detailed**                     **Schematic**

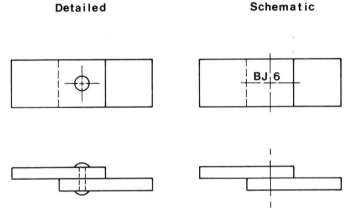

FIGURE 8-13   How to draw rivets.

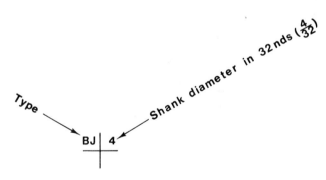

FIGURE 8-14   Definition of rivet notations.

FIGURE 8-15   How to call out a row of rivets.

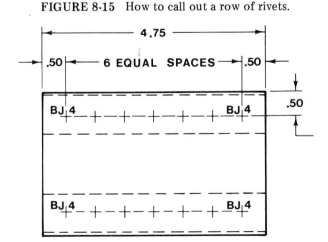

## SPOTWELDS

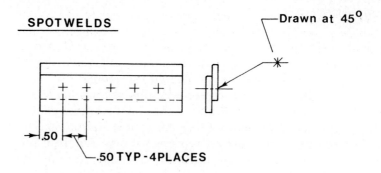

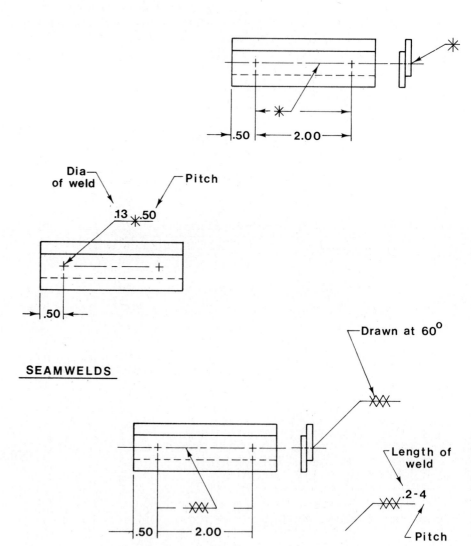

## SEAMWELDS

**FIGURE 8-16**   Spotwelds and seamwelds.

Spot welds are shown on a drawing using the symbols presented in Figure 8-16. They may be represented by either of the symbols shown. The dimensions for the symbol sizes are guidelines only and need not be followed strictly.

Spot welds are sometimes more advantageous in a design because they do not require any space, whereas a rivet requires space for its head and butt ends. Rivets, on the other hand, may be drilled out and replaced, whereas a spot weld must be rewelded. Rewelding usually requires almost complete disassembly of a chassis.

The desired diameter of a spot weld may be indicated by lettering in the diameter size to the left of the circle on the weld symbol. The distance between spot welds, measured from center to center and called the *pitch* of the spot welds, may be indicated by lettering-in the distance value to the right of the circle on the symbol (see Figure 8-16).

## PROBLEMS

8-1   Draw and dimension the flat pattern for the shapes defined in Figure P8-1.

### FIGURE P8-1

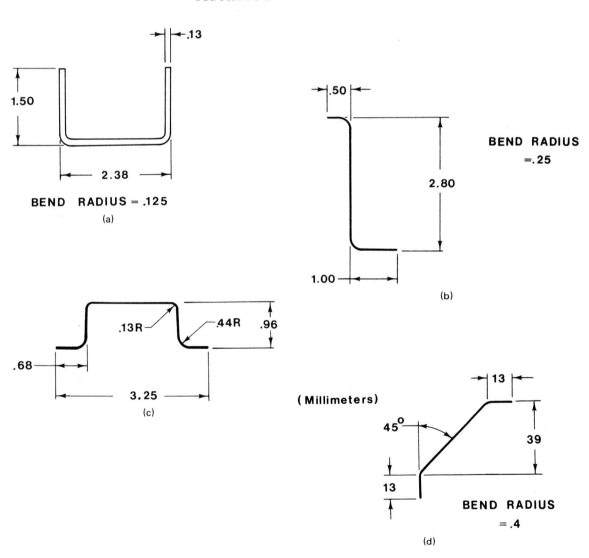

8-2    Dimension the chassis surface shown in Figure P8-2 using:
a.  The coordinate system
b.  The baseline system
Each block on the grid pattern equals 0.125 inch.

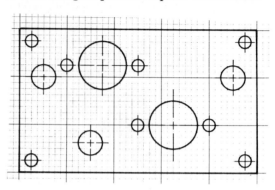

FIGURE P8-2

8-3    Dimension the chassis surface shown in Figure P8-3 using:
a.  The coordinate system
b.  The baseline system
Each block on the grid pattern equals 0.20 inch.

8-4    Dimension the chassis surface shown in Figure P8-4 using:
a.  The coordinate system
b.  The baseline system
Each block on the grid pattern equals 0.20 inch.

8-5    Draw and dimension a flat pattern for the chassis shape shown in Figure P8-5. The holes are to be drilled before the material is bent. Each block on the grid pattern equals 0.20 inch.

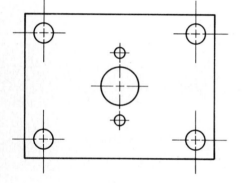

FIGURE P8-3

FIGURE P8-4

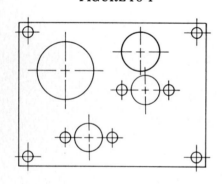

ALL DIMENSIONS TO THE NEAREST .10

FIGURE P8-5

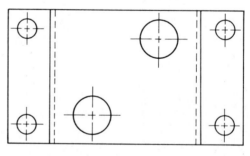

MATL
.13 THK

ALL  INSIDE  BEND
RADII = .10

**8-6**   Redraw the figure shown in Figure P8-6 and identify the fasteners
as:

a.   BJ6 rivets

b.   $^{3}/_{16}$-inch-diameter spot welds

Remove all fasteners and use a seam weld.

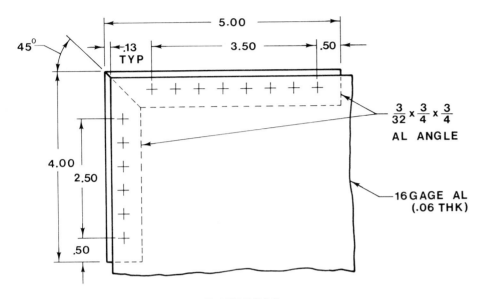

**FIGURE P8-6**

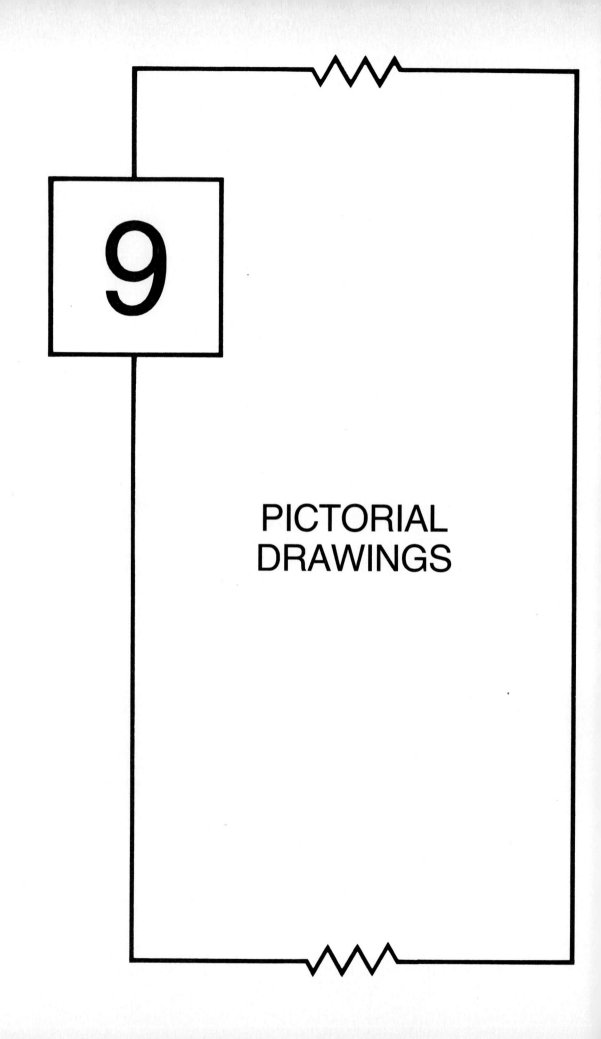

# 9

## PICTORIAL DRAWINGS

## 9-1 INTRODUCTION

Electrical and electronic drafters are often asked to prepare a pictorial drawing of the project on which they are working. These pictorial drawings may be used as part of a sales presentation to help customers better understand what they are purchasing, or they may be used in assembly or maintenance instructions to make it easier for crafts-persons and mechanics to understand the instructions.

There are many different types of pictorial drawings, but the isometric and oblique types are most often used. Figure 9-1 illustrates the two types of drawings.

Isometric drawings are the more visually correct of the two, but are more difficult to draw. Oblique views are easier to draw, but present a minimum pictorial representation.

## 9-2 ISOMETRIC DRAWINGS

*Isometric drawings* are pictorial drawings based on an isometric axis: that is, two 30° lines and one vertical line (see Figure 9-2). Customarily, the left isometric plane is used to show the front view of the object, the right plane the right-side view, and the top plane the top view. Figure 9-3 illustrates this convention.

To prepare an isometric drawing, use the following procedure, which is illustrated in Figure 9-4:

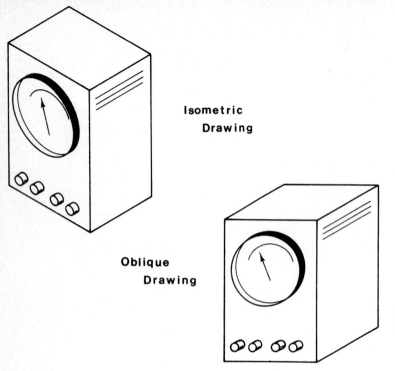

Isometric
Drawing

Oblique
Drawing

FIGURE 9-1 Isometric and oblique drawings of a meter.

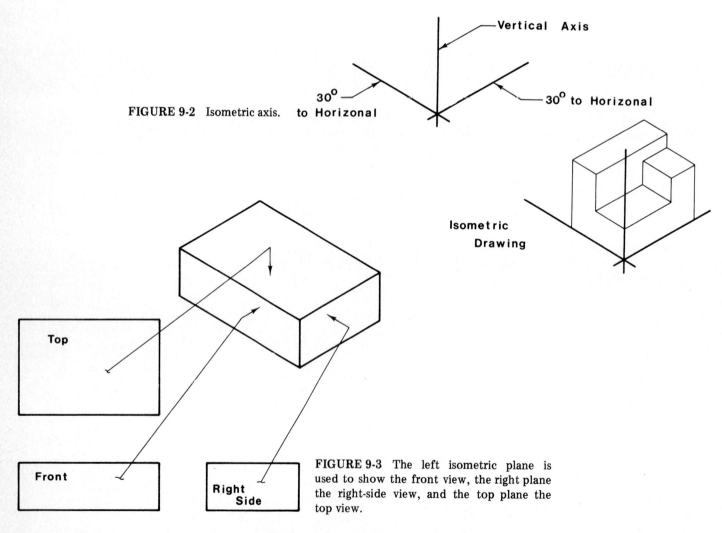

Isometric Axis

Vertical Axis

30° to Horizonal

30° to Horizonal

FIGURE 9-2 Isometric axis.

Isometric
Drawing

Top

Front

Right
Side

FIGURE 9-3 The left isometric plane is used to show the front view, the right plane the right-side view, and the top plane the top view.

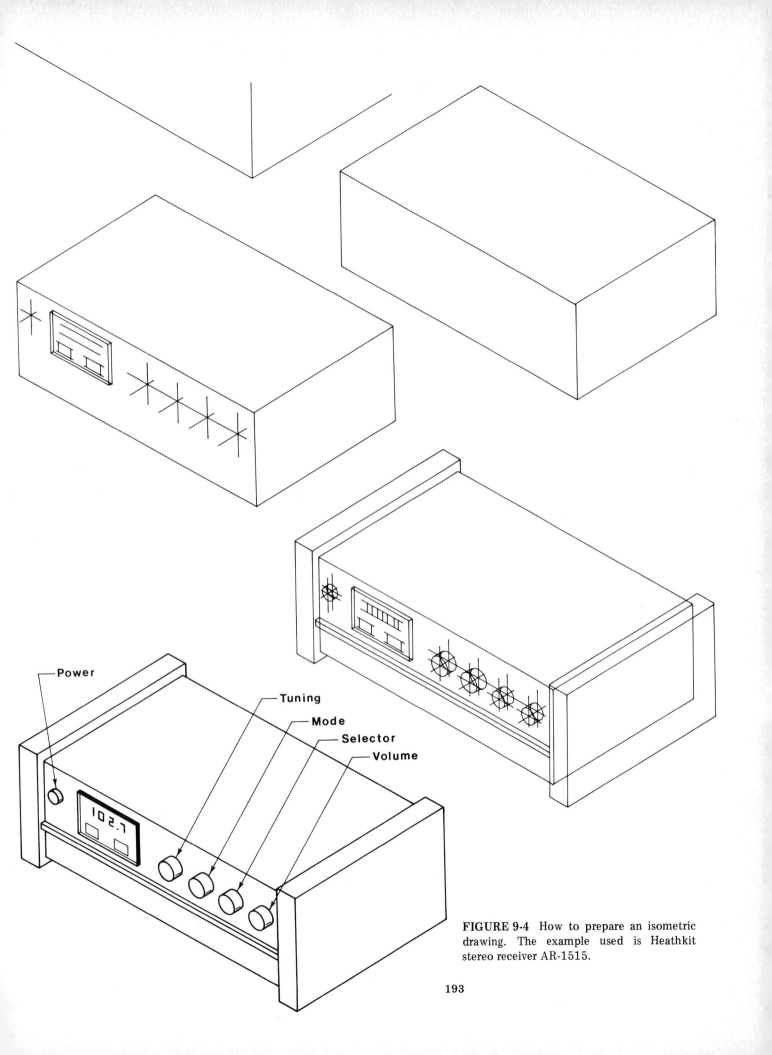

**Power**

**Tuning**

**Mode**

**Selector**

**Volume**

**FIGURE 9-4** How to prepare an isometric drawing. The example used is Heathkit stereo receiver AR-1515.

1. Lay out an isometric axis.

2. Draw the length, depth, and width of the object as shown.

3. Add, using very light construction lines, the preliminary detail of the object.

4. Add the final detail to the drawing.

5. Erase any excess lines and draw, using heavy black lines, the final shape of the object. The final drawing may be traced if the layout created in step 4 cannot be erased cleanly.

One shape that usually causes difficulty for beginners when preparing isometric drawings is a circle. This is because circles are not drawn as circles but as ellipses. Figure 9-5 illustrates the visual distortion that circles create when drawn on isometric drawings and shows the visually correct picture ellipses create.

Correctly shaped ellipses can easily be drawn on isometric drawings by using an isometric ellipse template. Figure 9-6 pictures an isometric ellipse template and demonstrates how the template is used. For ellipses in the left-hand plane, rotate the template as shown so that the edge labeled LEFT-HAND PLANE rests on the T-square. For ellipses in the right plane, use the edge labeled RIGHT-HAND PLANE as the base, and for ellipses in the top plane, use the edge labeled HORIZONTAL PLANE as a base.

Several other templates are used for preparing isometric drawings. An isometric protractor is helpful when marking off dial indications or meter scales, as shown in Figure 9-7. A small ellipse template, pictured in Figure 9-8, is used to help draw very small ellipses which are not included on the isometric ellipse template discussed previously.

**FIGURE 9-5** Isometric drawing requires ellipses to represent holes. Circles will appear distorted.

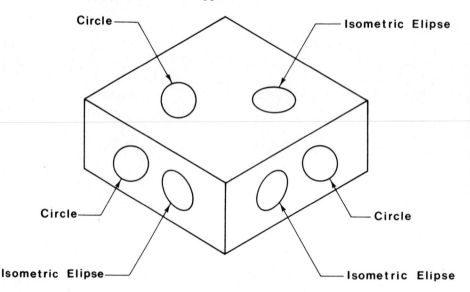

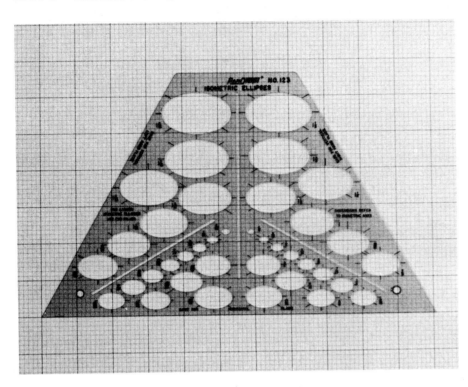

**FIGURE 9-6** How to use an isometric ellipse template.

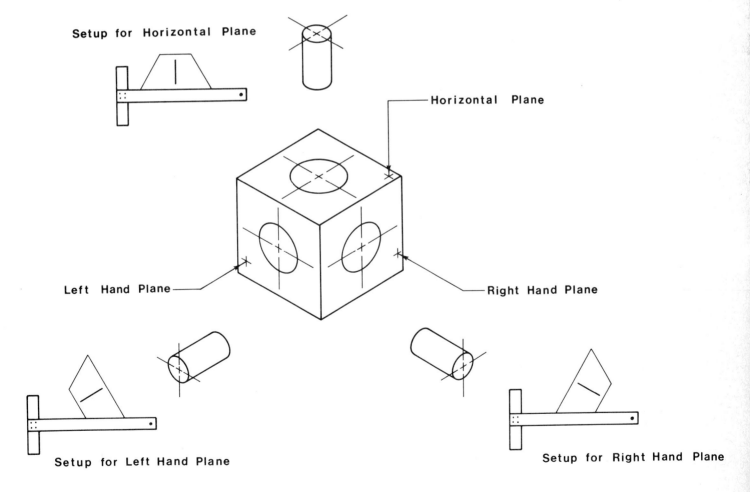

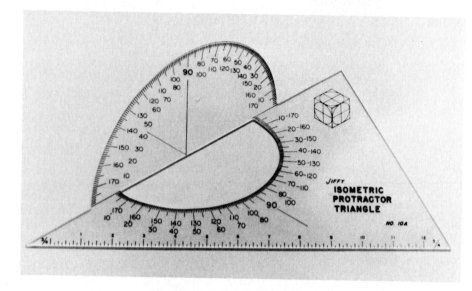

**FIGURE 9-7**   Isometric protractor.

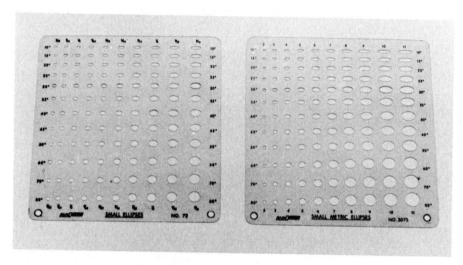

**FIGURE 9-8**   Small ellipse templates.

## 9-3   ISOMETRIC DRAWINGS OF CIRCUIT BOARDS

Figure 9-9 shows an isometric drawing of a circuit board. It is actually a composite of several smaller isometric drawings: five isometric drawings of resistors, one isometric drawing of a capacitor, two isometric drawings of transistors, and a circuit board, which are all combined to form the finished isometric drawing.

Figure 9-9 was drawn using the information presented by the pictorial schematic diagram shown in Figure 9-10. For example purposes the wiring paths have been omitted. The procedure used to create the isometric drawing shown in Figure 9-9 from the schematic diagram shown in Figure 9-10 is as follows:

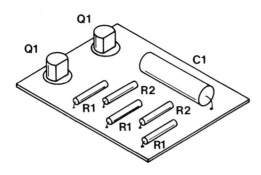

**FIGURE 9-9**   Isometric drawing of a circuit board.

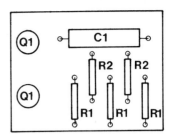

**FIGURE 9-10**   Pictorial schematic drawing from which Figure 9-9 was created.

1. Using very light lines, lay out an isometric drawing of the circuit board (see Figure 9-11).
2. Locate the components on the circuit board as shown.
3. Draw isometrically each individual component.
4. Trace, from the layout created in step 3, the finished isometric drawing. Drafters prefer to trace the finished drawing, since the layout created in step 3 is usually a maze of construction lines which is impossible to erase cleanly.

To summarize the procedure and to make it easier for you to prepare isometric drawings, Figure 9-12 has been included. Figure 9-12 is a pictorial listing of the components most often called for when creating isometric drawings. In each case, the step-by-step procedure used to create the finished isometric drawing is shown.

**FIGURE 9-11**   How to prepare an isometric drawing of a circuit board.

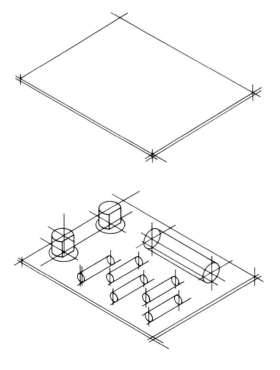

**FIGURE 9-12** Pictorial listing of various components often used for isometric electronic drawings.

**Circular Indicator**

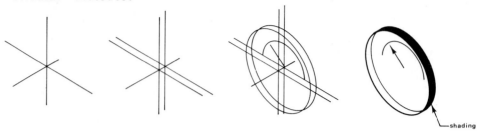

**Jack Receptacle**

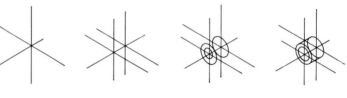

**Cylindrical Switch with Dial**

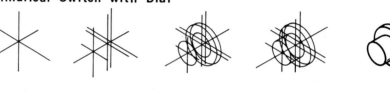

**Cylindrical Switch with Collar**

**Throw Switch**

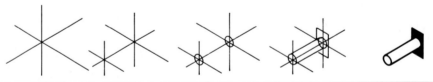

**Coil**

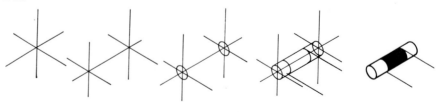

**Diode**

FIGURE 9-12   (cont.)

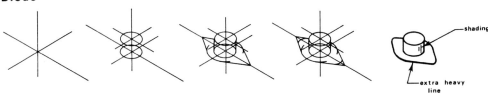

**Meter**

  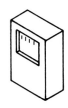

**Cylindrical Switch**

**Tapered Switch**

**Grooved Cylindrical Switch**

**Sliding Switch**

**FIGURE 9-12** (cont.)   **Protruding Square Switch**

**Recessed Square Switch**

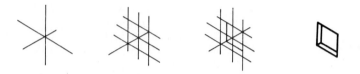

**Rectangular Indicator**

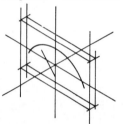
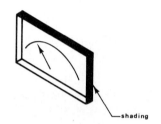

**Resistor or Fuse**

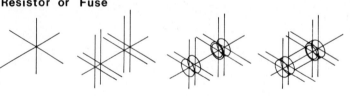

**Transformer**

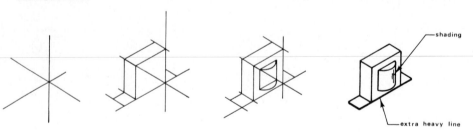

**Capacitor ( drawn freehand )**

**Resistor   or   Capacitor**                    FIGURE 9-12   (cont.)

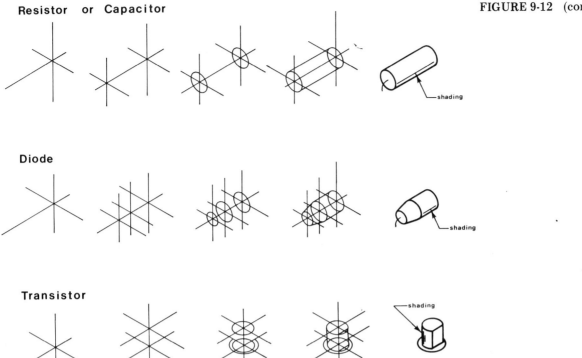

**Diode**

**Transistor**

## 9-4   OBLIQUE DRAWINGS

*Oblique drawings* are pictorial drawings which are based on an oblique axis: that is, a vertical line, a horizontal line, and a receding line (usually 30°) (see Figure 9-13). The principal advantage in preparing oblique drawings as opposed to other types of pictorial drawings is that the frontal plane contains a 90° angle. This means that round edges may be drawn as circles or arcs if they appear in the front plane or in a

**FIGURE 9-13**   Oblique axis.

**Oblique   Axis**

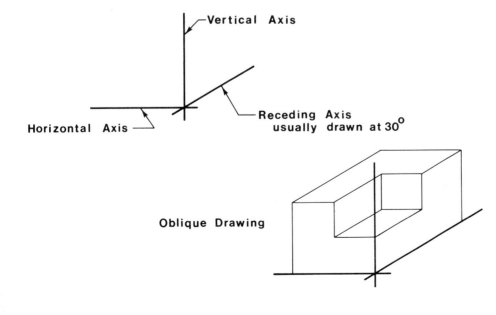

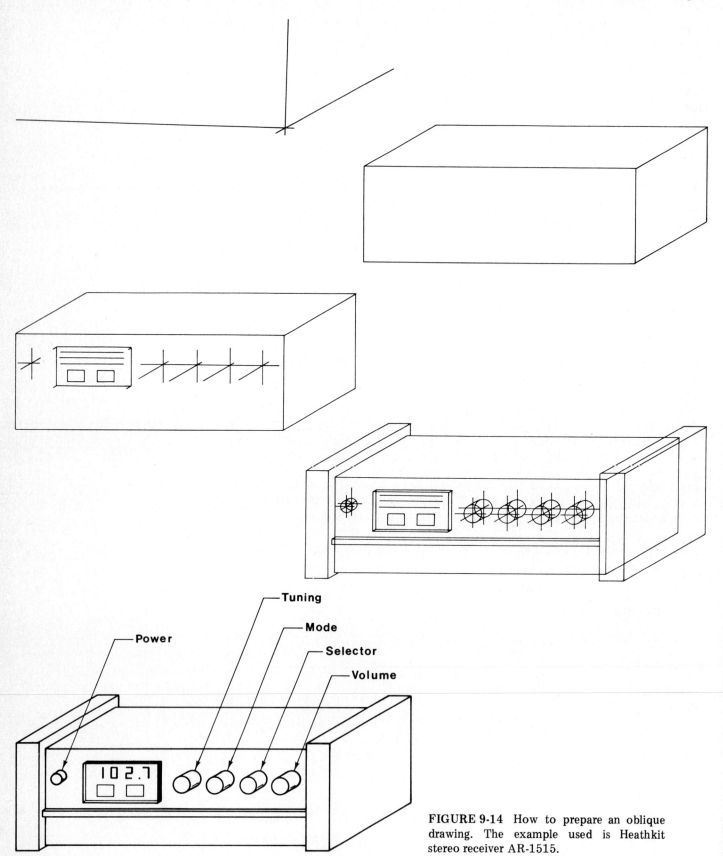

FIGURE 9-14  How to prepare an oblique drawing. The example used is Heathkit stereo receiver AR-1515.

plane parallel to the front plane. However, the advantage of being able to draw circles is applicable only to the frontal plane of oblique drawings. Receding planes require elliptical shapes for visual correctness just as did isometric drawings. This means that unless the object can be positioned so that the majority of circles and arcs are located in the frontal plane or in a plane parallel to the frontal plane, there is really no advantage to preparing an oblique drawing.

To prepare an oblique drawing, use the following procedure (illustrated in Figure 9-14):

1. Lay out an oblique axis.
2. Draw the length, depth, and width of the object as shown.
3. Add, using very light construction lines, the preliminary detail of the object.
4. Add the final detail to the drawing.
5. Erase any excess lines and draw, using heavy black lines, the final shape of the object. The final drawing may be traced if the layout created in step 4 cannot be erased cleanly.

Oblique drawings tend to make objects look deeper than they actually are. For example, compare the isometric and oblique drawings of Figure 9-1. The oblique drawing appears to be larger, when in fact, both drawings are drawn to the same scale (check for yourself). To overcome this visual distortion, oblique drawings are often drawn as *cabinet projections*.

Cabinet projections are drawn exactly the same as regular oblique drawings (which are called *cavalier projections*) except that dimensions along the receding axis are drawn at half size, that is, at ½ their scale values. Figure 9-15 illustrates. Note, in Figure 9-15, how much more proportioned the cabinet projection appears relative to the isometric drawing than does the full-size oblique drawing. The only dimensional differences between the three cubes in Figure 9-15 is the half-scale reduction of the receding axis values in the cabinet projection. Take a scale or pair of dividers and compare the objects of Figure 9-15 for yourself.

**FIGURE 9-15**  Comparison between a cavalier and cabinet projection.

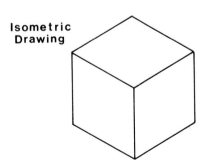

Isometric Drawing

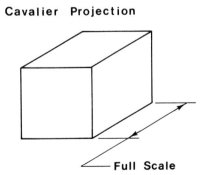

Cavalier Projection

— Full Scale

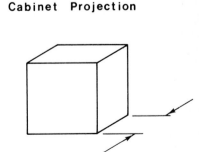

Cabinet Projection

— Half Scale

## 9-5  OBLIQUE DRAWINGS
## OF CIRCUIT BOARDS

Oblique drawings may be made of circuit boards using the same procedure outlined in Section 9-3 for isometric drawings. The only difference is that when creating an oblique drawing an oblique axis must be used.

Figure 9-12, which is set up to help create isometric drawings, may be modified to help create oblique drawings by changing the first step in each drawing outline from an isometric axis to an oblique axis. Figure 9-16 illustrates this change as applied to the drawing of a diode and a meter with a recessed face. Figure 9-17 is an example of an oblique drawing of a circuit board.

**Diode**

**Meter**

FIGURE 9-16  Oblique drawings of components. The procedure used to create the drawings is the same as was used for Figure 9-12. The only difference is the initial axis setup.

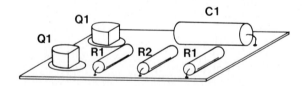

FIGURE 9-17  Oblique drawing of a circuit board.

## PROBLEMS

9-1  Prepare either an isometric or oblique drawing, as assigned by your instructor, of the Realistic components shown in Figure P9-1 (photographs are courtesy of Tandy Corp., Radio Shack). All dimensions are to be by eye, but try to keep the general proportions as pictured. The microphones and mounting brackets may be omitted. Use a large sheet of paper for Figure P9-1(d).

(a)

(b)

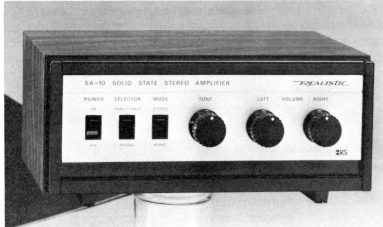

(c)

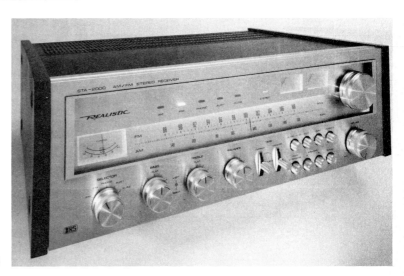

(d)

**9-2**  Prepare an isometric drawing of the component side of the PC board shown in Figure P9-2.

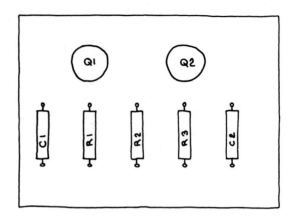

TRANSISTORS = .80 DIA. × .35 HIGH

RESISTORS & CAPACITORS = .30 DIA. × 1.00 LONG

BOARD SIZE = 4.00 × 5.00

**FIGURE P9-2**

**9-3**  Prepare an isometric drawing of the meter box sketched in Figure P9-3.

**FIGURE P9-3**

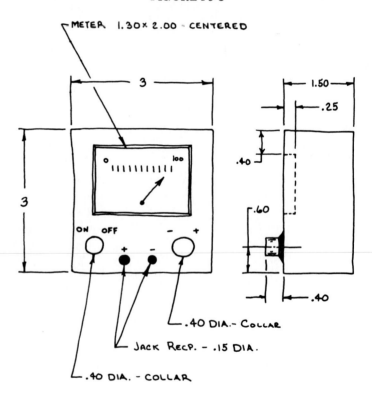

METER BOX

**9-4**  Prepare an isometric drawing of an amplifier that measures 4 ×
12 × 12 inches. Design the front panel so that it includes an off-
on pushbutton switch; a volume control; a tone switch; a speaker
control switch labeled *A*, *B*, *A* + *B*; a balance control switch; a
switch marker PHONO, TAPE, MIC, MISC; and an earphone jack
plug.

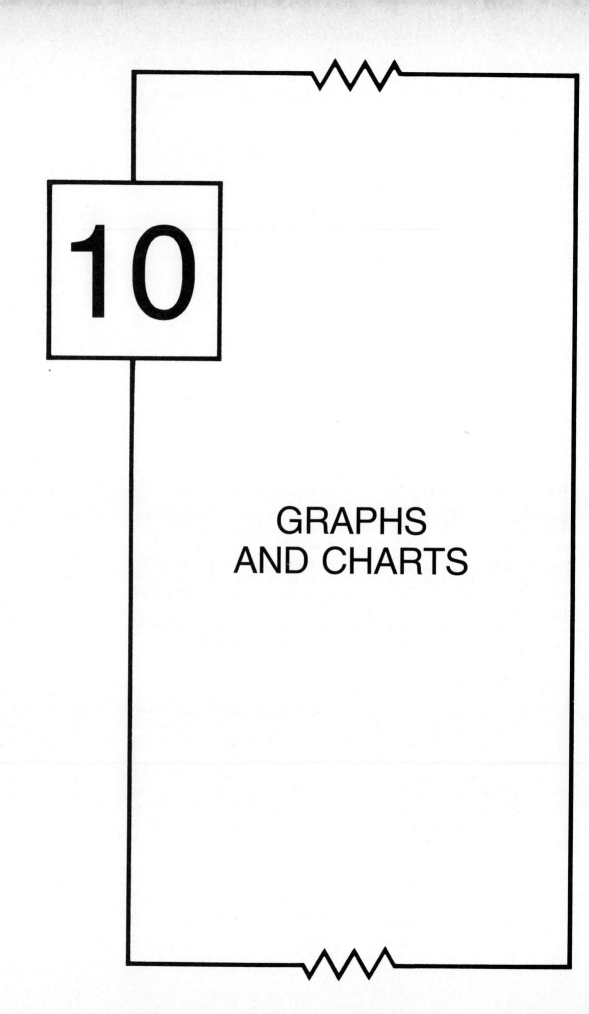

# 10

# GRAPHS
# AND CHARTS

## 10-1 INTRODUCTION

This chapter explains how to create graphs and charts. Drafters are often asked to convert lists of raw data into graph or chart form so that the data may be more easily understood. Most engineering test data, for example, originally appear in computer printout form—long, long lists of numbers. To read and analyze these lists is tedious work. If, however, the same data are presented in graph form, it is much easier to understand.

This chapter covers pie charts, bar charts, and curve plotting. Logarithmic and semilogarithmic grids and scales are also included.

## 10-2 PIE CHARTS

*Pie charts* are used to illustrate the relative sizes of various component parts of a total quantity. The name pie chart comes from its shape, which looks like an overview of a pie which has been sliced into different-sized pieces. Figure 10-1 illustrates a pie chart.

To demonstrate how to create pie charts, consider the following problem (Figure 10-2 illustrates). You are asked to draw a pie chart that will illustrate the relative numbers of persons serving the Army, Navy, Air Force, and Marine Corps in 1975. You are given the following figures:

| Military personnel, 1975 | |
| --- | --- |
| Army | 781,316 |
| Navy | 549,400 |
| Air Force | 608,337 |
| Marines | 192,200 |

**Production of Electricity in U.S. by Source — 1975**

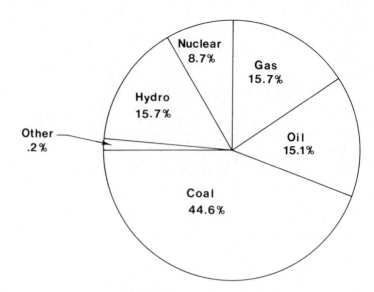

FIGURE 10-1   Example of a pie chart.

FIGURE 10-2   How to prepare a pie chart. The degree values are based on calculations.

**Persons Serving in the Army, Navy, Air Force, & Marines 1975**

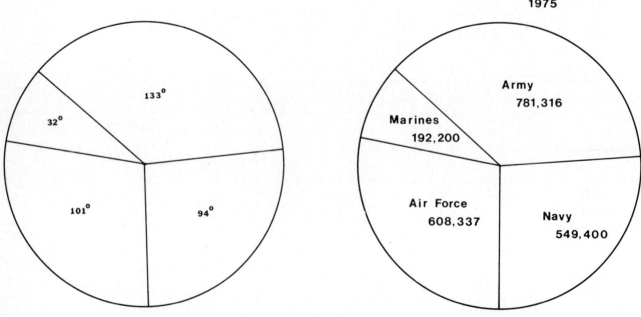

To convert these figures into a pie chart:

1. Calculate the percentage value of each figure. This is done by first finding the total value of all the figures and then dividing each figure into the total.

$$
\begin{aligned}
781{,}316 \\
549{,}400 \\
608{,}337 \\
\underline{192{,}200} \\
\hline
2{,}131{,}253
\end{aligned}
\quad \text{Total value}
$$

Army    $\dfrac{781{,}316}{2{,}131{,}253} = 0.37$ or 37%

Navy    $\dfrac{549{,}400}{2{,}131{,}253} = 0.26$ or 26%

Air Force    $\dfrac{608{,}337}{2{,}131{,}253} = 0.28$ or 28%

Marines    $\dfrac{192{,}200}{2{,}131{,}253} = 0.09$ or 9%

2. Convert the percentages calculated in step 1 into the equivalent percentage of a circle. This is done by multiplying the percentage values by $360°$:

Army    37% of $360° = 0.37(360°) = 133°$

Navy    26% of $360° = 0.26(360°) = 94°$

Air Force    28% of $360° = 0.28(360°) = 101°$

Marines    9% of $360° = 0.09(360°) = 32°$

These values are in terms of degrees, and can be measured using a protractor.

3. Lay out a circle about 6 inches in diameter and mark off the degree values calculated in step 2.

4. Label each sector as shown and add a title for the chart. All lettering should be neat and easy to read. If desired, the different sectors may be shaded differently to help distinguish them from each other.

## 10-3 BAR CHARTS

*Bar charts* are used to demonstrate the differences between fixed quantities. They derive their name from the fact that they express values in terms of bar-shaped figures, as shown in Figure 10-3.

It is important that bar charts be drawn with a properly proportioned scale. Scales that are too small are difficult to read and may not show the differences between values clearly, while scales that are too large may not fit the paper.

Because bar charts usually involve a comparison of fixed values,

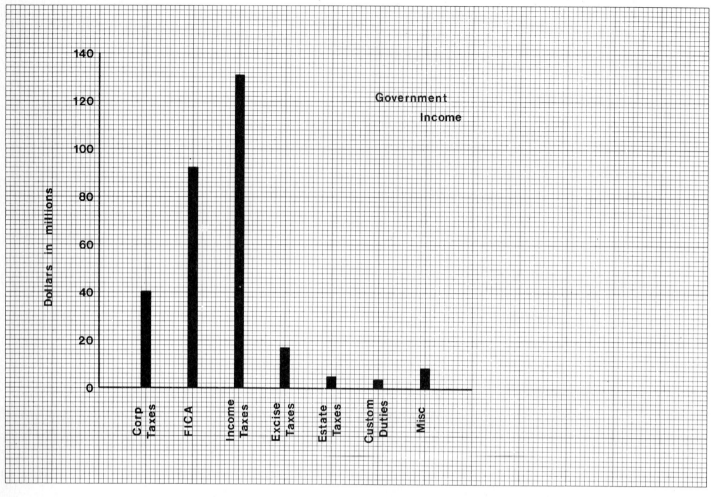

**FIGURE 10-3** Example of a bar chart.

the values are sometimes lettered in just above the top of the bar. In this way the viewer gets not only a visual comparison but also the numerical values so that further comparison may be made if desired. This information could be derived from the bar chart scale, but it is usually very helpful to the reader to have the values clearly stated.

Consider the following problem (Figure 10-4 illustrates). You are asked to create a bar chart that compares the number of home runs hit by the National League's leading home run hitters from 1920 to 1924. The number of home runs hit was:

| | | |
|---|---|---|
| 1920 | Cy Williams, Philadelphia | 15 |
| 1921 | George Kelly, New York | 23 |
| 1922 | Rogers Hornsby, St. Louis | 42 |
| 1923 | Cy Williams, Philadelphia | 41 |
| 1924 | Jacques Fourmier, Brooklyn | 27 |

To convert these figures into a bar chart:

1. Choose a scale. In this case we let each block on the vertical axis represent 1 home run, or 10 home runs to the inch. Whenever pos-

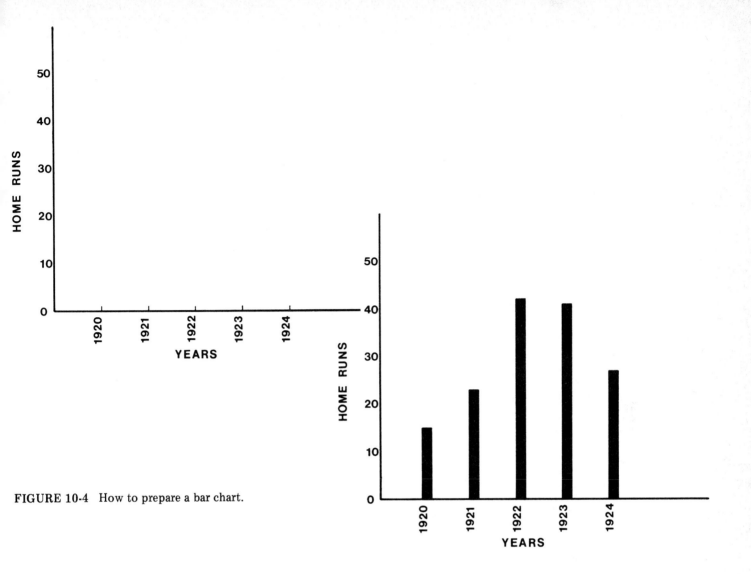

FIGURE 10-4  How to prepare a bar chart.

National League's
Home Run Leaders
1920 - 1924

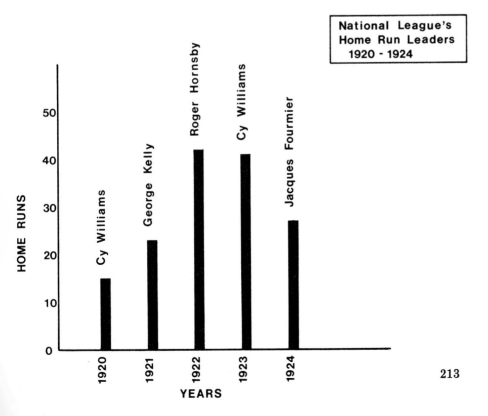

213

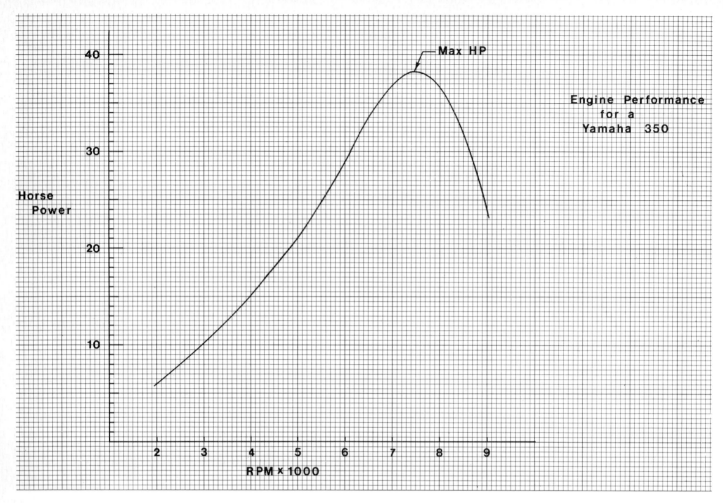

**FIGURE 10-5** Example of a curve that shows the relationship between two or more variables, in this case horsepower versus rpm.

sible, choose a scale that is directly related to 10, as most people are used to thinking in terms of a base 10.

2. Define the horizontal and vertical axes.
3. Draw in the bars.
4. Label each bar and add the title and any other necessary information.

It is not necessary to have the horizontal axis represent a value of 0. If there are extremely large values to work with, the length of the bars may be reduced by assigning a value higher than 0 to the horizontal axis.

## 10-4 CURVE PLOTTING

Curve plotting is a term used to describe the creation of graphs that are based on data that are a function of two or more variables (Figure 10-5 illustrates). Curve plotting does not produce a comparison between fixed values as do pie charts and bar graphs, but instead shows the relationship between two or more different variables.

For example, in Figure 10-5, we see that the horizontal axis is a measure of engine speed (RPM) and the vertical axis is a measure of horsepower (HP). Both engine speed and horsepower are variables. The shape of the curve shows the relationship between the two variables. The curve slopes upward, meaning that as values of engine speed increase, values of horsepower also increase, up to the point of maximum horsepower. After this point, as the values of engine speed increase, the values of horsepower decrease—the curve slopes downward. This means that after we pass the point of maximum horsepower, increasing the engine speed actually reduces the amount of horsepower the engine generates.

Four areas often cause trouble when plotting curves: choosing a scale, identifying the curves properly, drawing a smooth curve between the data points, and choosing the proper grid background (sometimes patterns other than squares are used). Each area will be considered separately.

## 10-5    CHOOSING A SCALE

A scale must be chosen so that it will be easy to read. It is tempting to pick a scale which is mathematically convenient—that is, the scale is easy to fit to the data, but such scales are usually very difficult to read and work with. For example, consider the two curves plotted in Figure 10-6. Both represent the same data, but the curve on the left uses an odd scale and the curve on the right uses a scale based on 10. What are the $x$ and $y$ values of point $A$? If we use the curve on the right we can easily find the answer $x = 8.0$, $y = 13.3$, but we see how much more difficult it is to derive the same answer from the left curve.

A scale must also consider the size of the paper on which the

**FIGURE 10-6**   Care must be taken when choosing a scale. What is the value of point $A$?

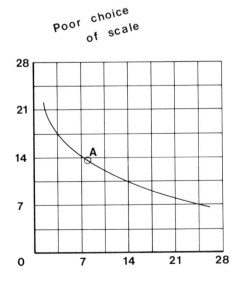

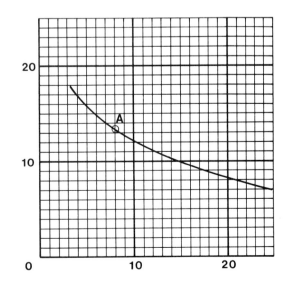

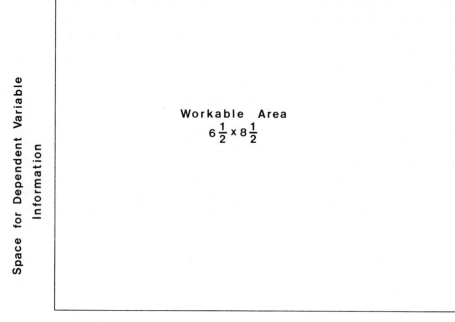

Space for Independent Variable
Information

**FIGURE 10-7**  Always leave room for identifying the axis.
This means that the area for curve plotting is actually less
than the total grid pattern.

curves are to be drawn. Most 8½ by 11 inch sheets of graph paper have
a grid pattern which is 8 by 10 inches (some types of graph paper have
even smaller grids). The workable area of the paper is further reduced
when the space needed to define the horizontal and vertical axes is
taken into account. This means that the final workable area—the area
in which we can draw curves—for an 8½ by 11 inch sheet of paper is
only about 7 by 9 inches at most (see Figure 10-7). To choose a scale
which is both easy to read and will fit the paper, consider the following
problem.

Plot the performance curve for a Texas Instruments SN52506 dual
differential comparator with strobe as a function of input offset current
versus the free air temperature. The test data are as follows:

| Free air temperature (°C) | Input offset current ($\mu$A) |
|---|---|
| −50 | 2.09 |
| −25 | 1.40 |
| 0 | 0.94 |
| 25 | 0.67 |
| 50 | 0.53 |
| 75 | 0.43 |
| 100 | 0.36 |

We first notice that the free air values are equally spaced and that
at least six spaces are needed. We also realize that the free air temperature values are the independent variables (the temperature of air does

not depend on the values of the comparator) and should be plotted along the horizontal axis.

If we let each inch represent 25°C of free air temperature, we need at least 6 inches along the horizontal axis. Normally, 25 units per inch would not be a good choice, but because we have evenly spaced values it is acceptable in this situation.

The input offset current values vary from 0.36 to 2.09 $\mu$A. If we let each inch equal 1.00 $\mu$A, we would need a little over 2 inches along the vertical axis. However, this would result in an almost flat curve which would not clearly represent the relationship between the variables. By doubling the scale so that two blocks equals 0.1 $\mu$A, the curve will require a little over 4 inches on the vertical scale.

It is customary to extend curves beyond their furthest data points, so we need, using the scales determined above, approximately a 5 × 7 inch area for the curve. This is well within the 6½ × 8½ inch limits outlined in Figure 10-7. Figure 10-8 illustrates how the final curve was plotted and labeled.

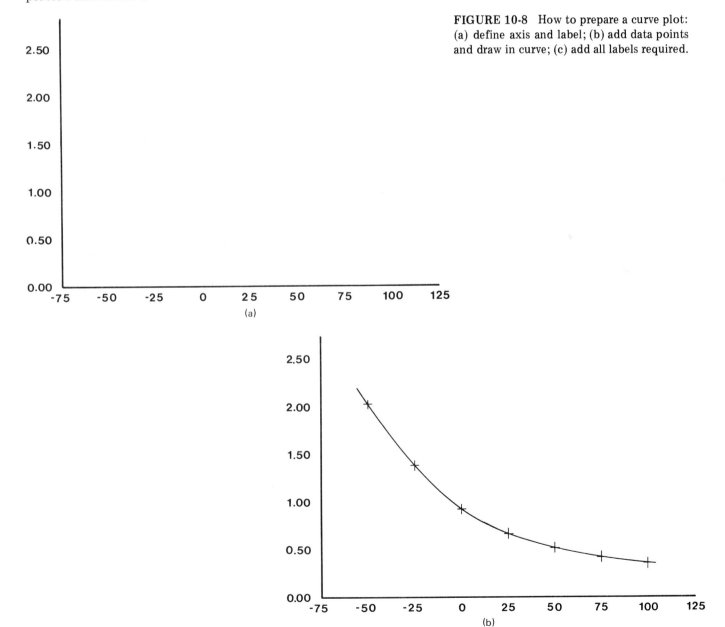

**FIGURE 10-8**  How to prepare a curve plot: (a) define axis and label; (b) add data points and draw in curve; (c) add all labels required.

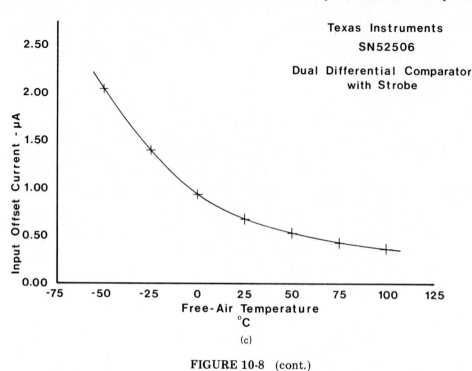

FIGURE 10-8   (cont.)

Figure 10-9 plots age versus height for males and females. Because the curves are almost identical up until age 12, a large vertical scale is required to bring out the small differences. Note how small lines were added to both the horizontal and vertical axes to make it easier to understand the scale used.

Note how in all curve plots the final graph includes completely defined horizontal and vertical scales, including labels, a title, a definition of data points and line patterns, any notes required, and labels for all curves.

## 10-6   IDENTIFYING CURVES

When two or more curves are drawn on the same graph, it is important that each be clearly identified. Several drawing techniques may be used to distinguish curves. A different color may be used for each curve, but most reproduction processes available to drafters do not reproduce color, so other methods must be used.

Figure 10-10 illustrates four different line patterns, and four different data point symbols, which can be used together to help distinguish curves. Note how curve 1 uses a small circle to indicate its data points and a broken-line pattern for the actual curve. Each of the other curves uses a different line pattern and data point symbol, all of which are defined in the accompanying legend. Each curve may be further labeled as shown, but this is an optional practice.

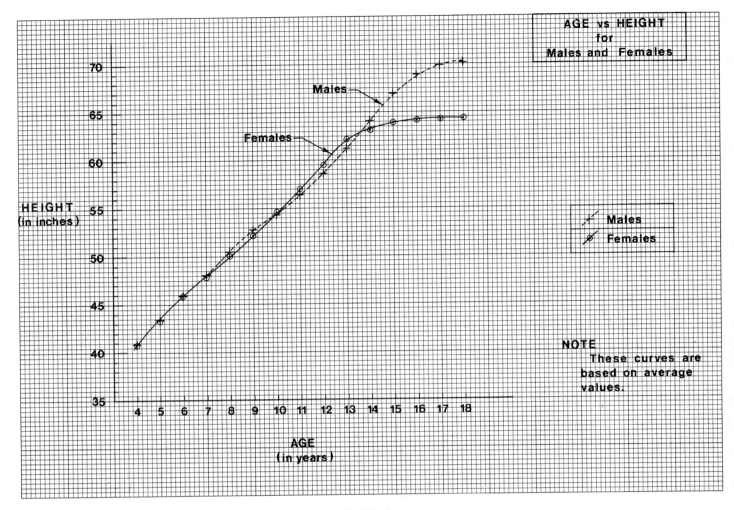

**FIGURE 10-9** Sample curve plot of age versus height for males and females.

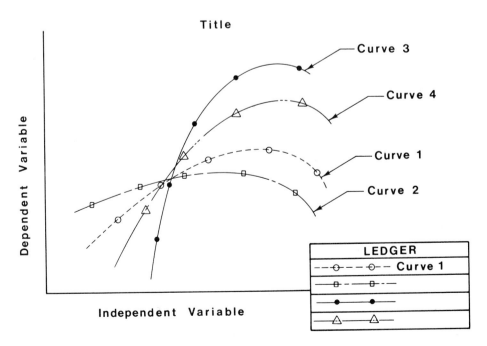

**FIGURE 10-10** When plotting different curves on the same axis, use different line patterns and different data point symbols to help the reader distinguish the curves.

## 10-7    DRAWING A SMOOTH CURVE

Once data points have been located on the graph paper, the problem remains of joining them with a smooth curve. Drafters create smooth curves by using plastic curves or adjustable curves as guides. Some examples of these devices are pictured in Figure 10-11.

Most students make the error of trying to connect too many points with one positioning of the curve. Figure 10-12 shows a series of points that are partially connected. The curve is in position to serve as a guide for joining *only* points 3 and 4—not 3, 4, and 5—even though all three seem to be aligned. To join point 5 using the curve position as shown would make it almost impossible to draw a smooth curve.

It is sometimes *not* necessary to connect all data points with a curve that goes through every data point. Curves that pass through every data point may be so wavy that they would be almost impossible to analyze. Further, in many cases, an exact analysis is not necessary since basic trends and patterns can be derived from an *average curve*.

**FIGURE 10-11** Some of the many different types of curves used by drafters to draw curves.

**FIGURE 10-12** How to draw a smooth curve between data points.

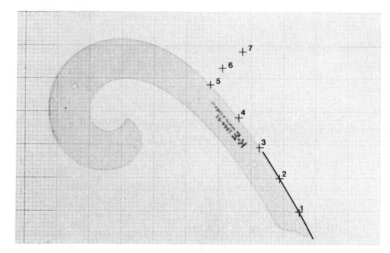

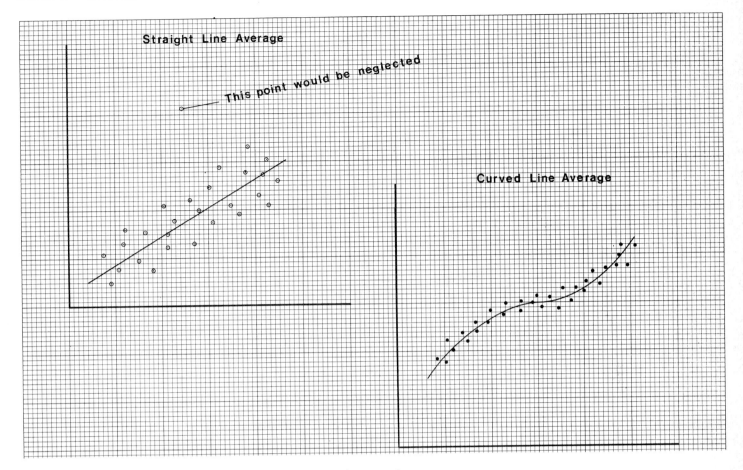

**FIGURE 10-13**  Examples of a straight-line and a curved-
line average.

Figure 10-13 shows two different types of average curves. The
straight-line average is created by drawing a straight line through the
data points so that approximately half the points are above the line and
half below the line. The distance between the points above the line
should be approximately equal to the distance between the line and the
points below the line. Curved-line approximations can be made in a
similar manner.

## 10-8   CHOOSING A GRID PATTERN

Up to now, we have considered only square grids, that is, patterns
whose lines are evenly spaced in both the horizontal and vertical directions. Many other patterns are used to plot data, including logarithmic,
semilogarithmic, and many different circular grids. Figure 10-14
pictures a sheet of graph paper printed with a polar grid pattern.

Figure 10-15 illustrates a logarithmic and a semilogarithmic grid
pattern. The logarithmic pattern is usually referred to as *log-log paper*
and the semilogarithmic pattern as *semilog paper*. Both grids are based
on the logarithmic scale, also illustrated in Figure 10-15.

Log-log and semilog grids are used because, for some data, they
make analysis easier. After drafters plot up a series of data, engineers
usually try to analyze it by trying to find the mathematical equation

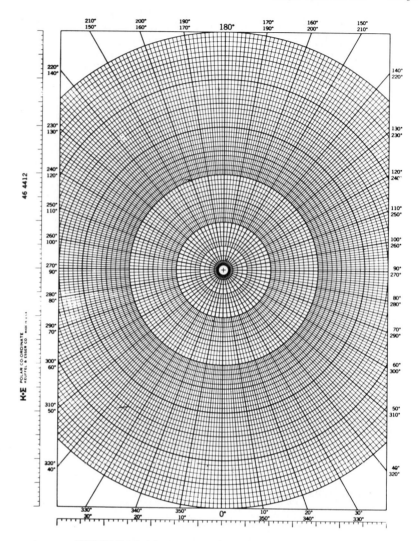

**FIGURE 10-14**  Sample sheet of polar graph paper.

that matches the curve. The work is, of course, greatly simplified if the data plot as a straight line, because the equation for any straight line is

$$y = mx + c$$

Some data that plot as a curve on a square grid pattern plot a straight line on a logarithmic grid, as Figure 10-16 illustrates.

The basic logarithmic grid can represent any multiple of 10 of the numbers 1 through 10. It can represent 1 through 10, 10 through 100, 100 through 1000, 0.1 through 1.0, and so on. For example, point A in Figure 10-17 can equal 0.034, 0.34, 3.40, 34, or 340 depending on which scale is used. Similarly, point B can equal 0.062, 0.62, 6.20, 62, or 620.

A unique feature of log grids is that they can *never* have a line with a value equal to 0.0. A line may be defined as 0.00001 or even smaller, but it can never be 0.0. This means that curves drawn on log grids can never cross, or even touch, the vertical axis.

Log scales may be placed one after another as shown in Figure 10-17 to increase the overall value range of the graph. It is good practice to label at least the beginning and end values of each log pattern used.

**Logarithmic Grid ( log–log )**            **Semilogarithmic Grid ( semi-log )**

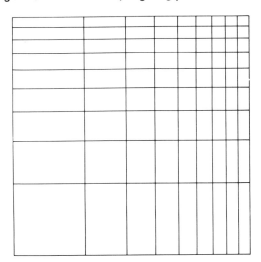
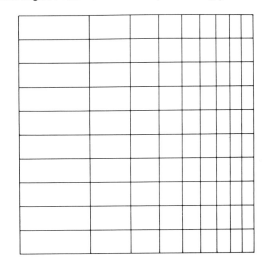

**Logarithmic Scale**

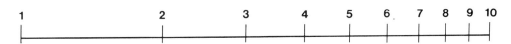

**FIGURE 10-15** Samples of a logarithmic and semilogarithmic grid pattern and a sample logarithmic scale.

**FIGURE 10-16** Sample of data points that plot as a straight line on a log-log pattern.

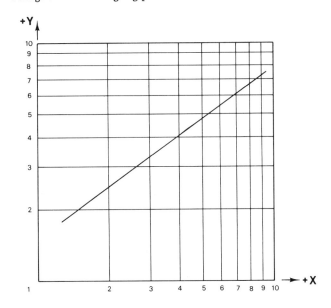

| X | Y |
|---|------|
| 2 | 2.52 |
| 3 | 3.36 |
| 4 | 4.10 |
| 5 | 4.75 |
| 6 | 5.46 |
| 7 | 6.08 |
| 8 | 6.62 |

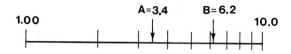

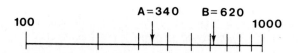

**FIGURE 10-17** How to read different log scales.

## PROBLEMS

**10-1** Figure P10-1 illustrates three different mathematical curves, that is, curves whose data points were obtained by substituting values into mathematical equations. In each example the $x$ variables are considered the independent variable and the $y$ variables are considered the dependent variable. Plot the following equations as assigned by your instructor. Include your calculation sheets. Always define the $x$ variable as the independent variable.

a.  $y = x$
b.  $y = 2x$
c.  $y = -\frac{1}{2}x$
d.  $y = 1.6x + 2$
e.  $4y = 2x - 1.50$
f.  $y = x^2$
g.  $y = x^2 + 2x$
h.  $y = \frac{1}{2}x^2 + 2x + 4$
i.  $y = 1.6x^2 - 2.2x - 0.7$
j.  $2y = -x^2 + 3x - 0.5$
k.  $xy = 4$

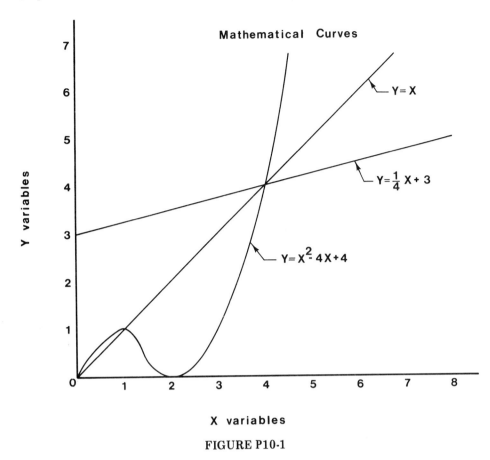

**X variables**

**FIGURE P10-1**

**10-2**   Draw a pie chart based on the following data:

World Power Consumption, 1976

| Country | Kilowatt-hours |
|---|---|
| United States | 1,999,688 |
| USSR | 1,038,000 |
| Japan | 432,757 |
| West Germany | 301,800 |
| Others | 2,480,755 |
| World total | 6,253,000 |

**10-3**   Draw a pie chart based on the following data:

U.S. Government Crime Report, 1977

| Murder | 20,510 |
|---|---|
| Forcible rape | 56,090 |
| Robbery | 464,970 |
| Aggravated assault | 484,710 |
| Burglary | 3,252,100 |
| Larceny-theft | 5,977,700 |
| Auto theft | 1,000,000 |
| Total | 11,256,000 |

**10-4**    Draw a bar chart based on the following data:

### Electricity Generated by Atomic Power

| Country | Output (MW) |
|---|---|
| United States | 28,236 |
| Great Britain | 4,258 |
| USSR | 4,521 |
| Japan | 3,718 |
| France | 2,886 |

**10-5**    Draw a bar chart based on the following data:

### Leading Lifetime Scores in Professional Football through 1977

| Name | Points |
|---|---|
| George Blanda | 2002 |
| Lou Groza | 1608 |
| Fred Cox | 1227 |
| Jim Bakken | 1171 |
| Jim Turner | 1147 |
| Gino Cappelletti | 1130 |
| Bruce Gossett | 1031 |

**10-6**    Draw a bar chart based on the following data:

### Leading Rebounders in NBA, 1975–1976

| Name | Team | Rebounds |
|---|---|---|
| Abdul-Jabbar | Los Angeles | 1383 |
| Cowens | Boston | 1246 |
| Unseld | Washington | 1036 |
| Silas | Boston | 1025 |
| Lacey | Kansas City | 1024 |

**10-7**    Plot the following data. Plot the years along the horizontal axis and the hourly earnings along the vertical axis.

### Increases in Factory Workers' Hourly Earnings

| Year | Hourly earnings |
|---|---|
| 1965 | 2.61 |
| 1966 | 2.72 |
| 1967 | 2.83 |
| 1968 | 3.01 |
| 1969 | 3.19 |
| 1970 | 3.36 |
| 1971 | 3.56 |
| 1972 | 3.81 |

**10-8**    Ohm's law defines the relationships among current, voltage, and resistance as

$$I = \frac{E}{R}$$

where $I$ = current (amperes)

      $E$ = voltage (volts)

      $R$ = resistance (ohms)

Plot a family of curves comparing amperes and volts at four different values of ohms. Plot the volts along the horizontal axis.

     The values to be plotted may be calculated by first assuming a constant value for ohms and then inserting various values for volts and solving the equation for amperes. For example, assume that the ohms value equals 2 ohms. If we solve the equations for voltage values of 1, 2, 3, 4, and 5 we have

$$I = \frac{1}{2} = 0.5$$

$$I = \frac{2}{2} = 1.0$$

$$I = \frac{3}{2} = 1.5$$

$$I = \frac{4}{2} = 2.0$$

$$I = \frac{5}{2} = 2.5$$

We can then plot the values:

| Volts | Amperes |
|-------|---------|
| 1 | 0.5 |
| 2 | 1.0 |
| 3 | 1.5 |
| 4 | 2.0 |
| 5 | 2.5 |

Label each curve as to its constant ohms value.

**10-9**    Plot the following equations on semilog paper.
     a.   $y = x$
     b.   $y = 3x + 2$
     c.   $y = x^2$
     d.   $y = x^2 + 1.5x - 4$
     e.   $y = e^x$
     f.   $y = e^{x+2}$
     g.   $y = e^x + 0.60$

**10-10**    Plot the following equations on log-log paper.
     a.   $y = x$
     b.   $y = -x^2 + 2.4$
     c.   $y = e^x$
     d.   $y = e^{x+2}$
     e.   $y = e^x + 2$

# 11

## RESIDENTIAL ELECTRICAL WIRING

## 11-1 INTRODUCTION

This chapter deals with residential electrical wiring drawings. We explain basic fundamentals of architectural drawing, define and show the application of the graphic symbols used on residential wiring drawings, and cover some of the basic design concepts required for residential wiring.

The information presented is in agreement with the standards set forth in the *National Electrical Code*® published by the National Fire Protection Association, 470 Atlantic Ave., Boston, MA 02110. Interested students may purchase copies of the Code by writing to the NFPA at the address above.

## 11-2 BASIC ARCHITECTURAL DRAFTING

The principal type of drawing used by architects is called a *floor plan*. A floor plan, as the name implies, is a drawing that defines the location of the various rooms, windows, doors, stairs, closets, hallways, and so on, for a given residence. Figure 11-1 shows a floor plan together with an elevation drawing of a small one-bedroom house.

The floor plan in Figure 11-1 has been labeled so that you can learn how to interpret architectural drawings. Note how doors, windows, and so on, are represented. All linework is as outlined in Figure 1-17 and all lettering is basically the same as shown in Figure 1-9, although it should be pointed out that architectural lettering is usually much more stylish than basic mechanical lettering.

The standard scale used for residential drawings is ¼ inch = 1

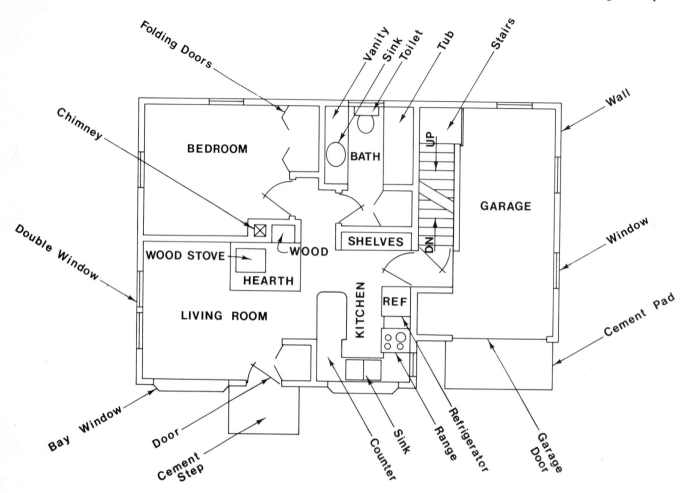

**FIGURE 11-1** Floor plan and elevation drawing of a small house. (The elevation drawing was prepared by Raymond Collard and the house design was created by Arthur Nelson, both of Wentworth Institute of Technology.)

foot; every ¼ inch on the drawing equals 1 foot on the house. Figure 11-2 illustrates a ¼-inch scale together with some sample measurements. The ¼ inch to the right of the 0 mark on the sample scale is graduated into 12 equal divisions so that measurements in inches can be made. Some scale manufacturers combine the ¼-inch scale, which reads from right to left, with a ⅛-inch scale, which reads left to right. The smaller numbers included in Figure 11-2 refer to values on a ⅛-inch scale.

## 11-3 ELECTRICAL SYMBOLS

Symbols are used on residential electrical wiring drawings to specify the type and location of the switches or outlets required. Figure 11-3 illustrates the most commonly used symbols and their meaning. The dimensions given are only approximations, as no national standard sizes have been agreed upon.

**FIGURE 11-1**   (cont.)

**FIGURE 11-2**   A ¼-inch scale with some sample readings.

**Readings are based on a scale of** $\frac{1''}{4}=1'$

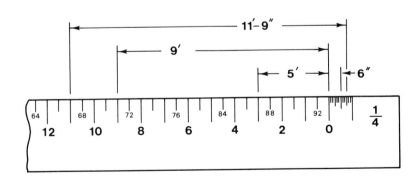

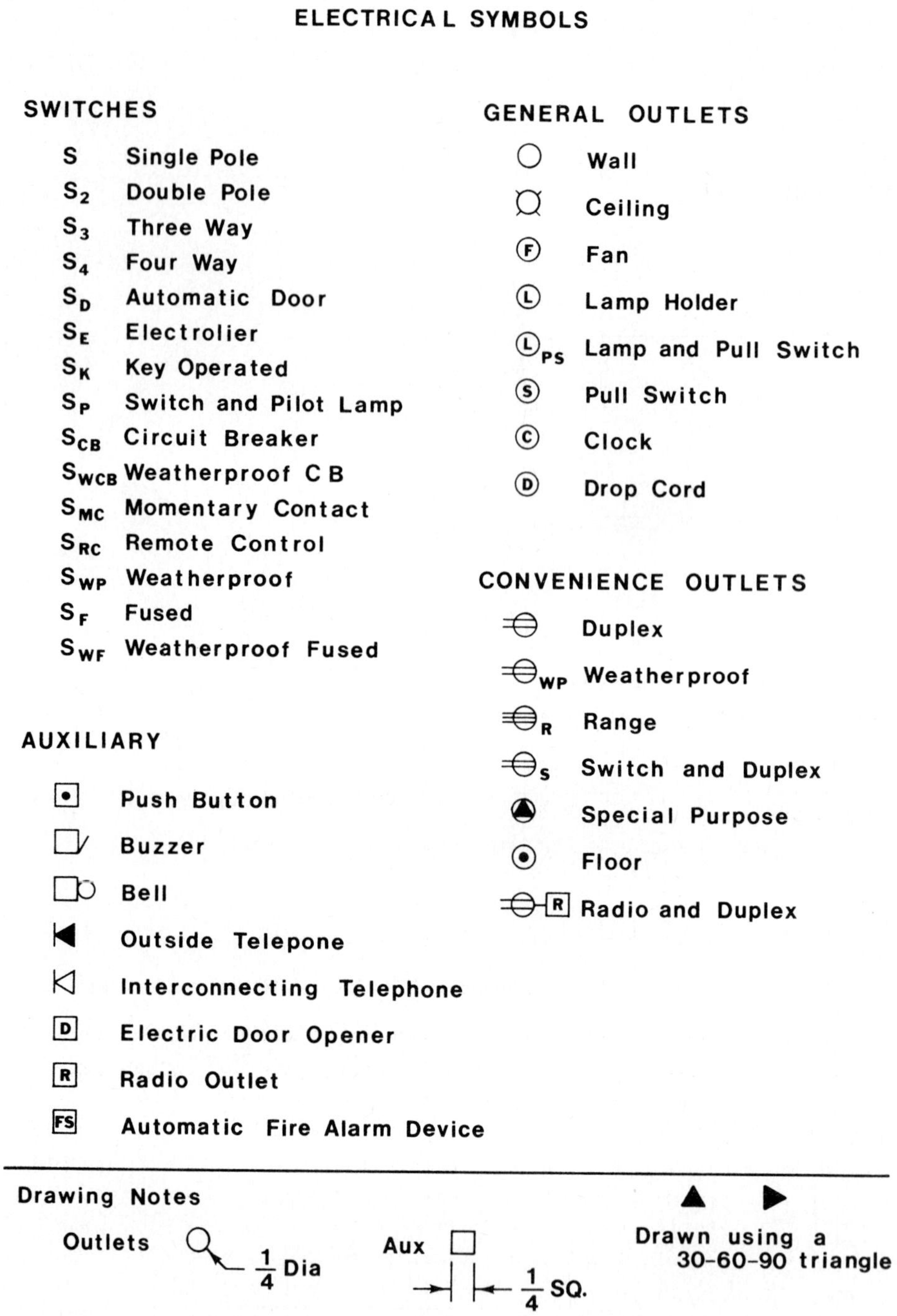

## ELECTRICAL SYMBOLS

### SWITCHES

| | |
|---|---|
| S | Single Pole |
| $S_2$ | Double Pole |
| $S_3$ | Three Way |
| $S_4$ | Four Way |
| $S_D$ | Automatic Door |
| $S_E$ | Electrolier |
| $S_K$ | Key Operated |
| $S_P$ | Switch and Pilot Lamp |
| $S_{CB}$ | Circuit Breaker |
| $S_{WCB}$ | Weatherproof C B |
| $S_{MC}$ | Momentary Contact |
| $S_{RC}$ | Remote Control |
| $S_{WP}$ | Weatherproof |
| $S_F$ | Fused |
| $S_{WF}$ | Weatherproof Fused |

### AUXILIARY

| | |
|---|---|
| ⊡ | Push Button |
| ▱ | Buzzer |
| ▭ | Bell |
| ◀ | Outside Telepone |
| ◁ | Interconnecting Telephone |
| D | Electric Door Opener |
| R | Radio Outlet |
| FS | Automatic Fire Alarm Device |

### GENERAL OUTLETS

| | |
|---|---|
| ○ | Wall |
| ◯ | Ceiling |
| Ⓕ | Fan |
| Ⓛ | Lamp Holder |
| Ⓛ$_{PS}$ | Lamp and Pull Switch |
| Ⓢ | Pull Switch |
| Ⓒ | Clock |
| Ⓓ | Drop Cord |

### CONVENIENCE OUTLETS

| | |
|---|---|
| ⊖ | Duplex |
| ⊖$_{WP}$ | Weatherproof |
| ⊖$_R$ | Range |
| ⊖$_S$ | Switch and Duplex |
| ⬤ | Special Purpose |
| ⊙ | Floor |
| ⊖—R | Radio and Duplex |

**Drawing Notes**

Outlets ○⟍ $\frac{1}{4}$ Dia     Aux ▢ →||← $\frac{1}{4}$ SQ.     ▲ ▶
Drawn using a
30–60–90 triangle

FIGURE 11-3 Electrical symbols for residential wiring drawings.

Figure 11-4 illustrates how these symbols may be added to a floor plan to create an electrical wiring drawing. In some cases, the electrical wiring drawing is created by making a blueprint of the floor plan and then drawing on the appropriate symbols with either a red or a yellow pencil.

Wiring paths are drawn as freehand hidden lines (a series of dashes; see Figure 1-17). They are drawn freehand so that they are easily dis-

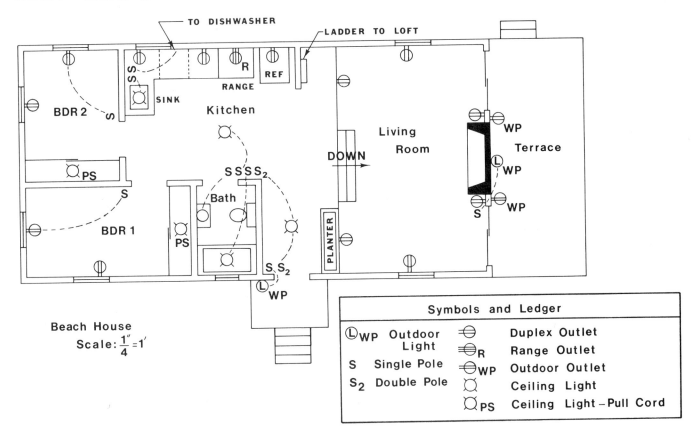

**FIGURE 11-4** Example of a residential wiring drawing.
The floor plan is for a summer beach house.

tinguishable from the floor plan lines. Note that the wire paths are specified only between switches and the fixtures they activate. It is not necessary to show all the wiring paths within a residence.

Wiring paths are omitted for two reasons. To show all the wiring would create a very cluttered, difficult-to-follow drawing which could lead to interpretation errors. It is also not necessary to show all the paths. Electricians need only be shown the location and type of fixture required and they will be responsible to ensure that the fixture is wired properly in accordance with the NEC and any local variances.

If you have trouble drawing neat freehand lines, use a french curve as a guide. The curve should help steady your hand.

Every residential electrical wiring drawing you create should include a symbol ledger which illustrates each symbol used on the drawing and defines its meaning. A sample symbol ledger is located on the drawings presented in Figure 11-4. The ledger may also be used to define any new symbols needed. Sometimes special equipment, such as smoke detectors, must be specified on a drawing and no standard symbol has been defined to cover the equipment. In such cases, a drafter may make up a symbol *provided* that it is clearly defined in the drawing ledger.

If a special or specific type of switch or outlet is required, it must also be defined on the drawing. This is usually done by printing the manufacturer's name and the part number of the fixture next to the appropriate symbol on the drawing. It may also be done by creating a special symbol and defining its meaning in the drawing ledger.

## 11-4  LOCATING SWITCHES AND OUTLETS

When choosing locations for electrical switches and outlets, it is important to consider how the switches and outlets will be used. For example, the switch shown in Figure 11-5 is not located properly. It is behind the door and would require a person entering the room to start to close the door in order to reach the switch. This is unacceptable, particularly if the room is dark.

Another error illustrated in Figure 11-5 concerns the stairs. There is no light located near or over the stairs, meaning that the stairs cannot be seen clearly at night.

The kitchen, shown in Figure 11-5 is also not lighted properly, even though the large fluorescent fixture would seem to be able to generate more than enough light. The problem here is one of shadows. A person working at the sink would have the light at his or her back, causing a shadow to fall over the sink. Even though the room is well lighted, the sink would be in darkness, making it inconvenient to use.

Shadows can also cause problems in the bathroom. In Figure 11-5 there is a light over the sink, but none over the tub. If a shower curtain were added (this would not necessarily appear on the floor plan) with the lighting setup shown, the light would be blocked out by the curtain and the person showering would be in greatly reduced light.

It is good practice, when preparing your drawing, to pretend that you are actually in the house on which you are working. Assume that it is dark, and mentally walk through the house checking that all switches and outlets that you have specified are located conveniently. For example, if you are working on a two-story house, and you are on the first floor, you will want a light over the stairs, a switch on the first floor to activate the light, and another switch at the top of the stairs to deactivate the light.

Lights can be arranged with two, three, or four separate activating switches. The symbols for double-pole, three-way, and four-way switches are shown in Figure 11-6.

To help you locate switches properly, the following checklist has been prepared. Check your drawings against the list to help prevent embarrassing errors.

1. When entering a room, is a light switch easily accessible?
2. For rooms with two or more entrances, are light switches located at each entrance?
3. Are all outside entrances lighted?
4. Are all hallways and stairs lighted with switches located at each end?

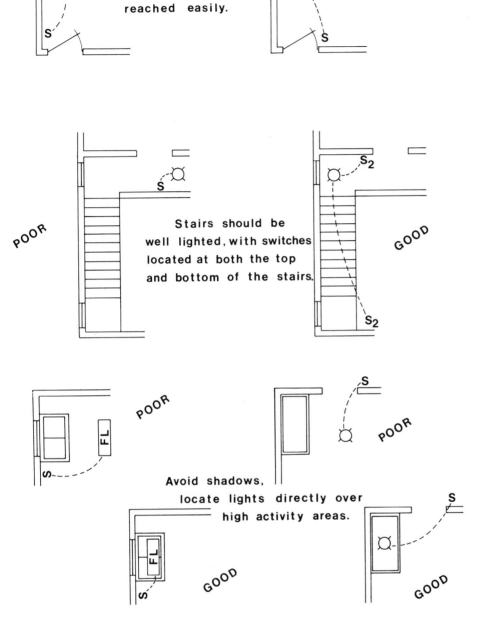

**FIGURE 11-5** Examples of good and poor electrical design.

5. Are all work areas (sinks, counters, etc.) lighted so as to prevent excessive shadows?

6. Are all special electrical requirements (range, dishwasher, dryer, air conditioners, etc.) taken into account?

7. Are all outside fixtures waterproof?

**A single pole switch which
is wired to a duplex outlet; the
switch activates or deactivates
the outlet.**

Only REPRESENTS
wiring path, actual wiring
path is in the wall.

**Two double pole switches wired
to a single ceiling outlet; either
switch can activate or deactivate
the outlet.**

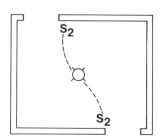

**Three three-way switches all wired
to a single ceiling outlet; any of three
switches can be used to activate or de-
active the outlet.**

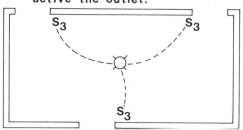

**FIGURE 11-6** How to indicate double-pole, three-way,
and four-way switches on a drawing.

## 11-5   ELECTRICAL TERMINOLOGY

Several terms commonly used in residential electrical wiring are listed
below. Each is defined and explained. Figure 11-7 illustrates some of
the terms.

*Branch Circuit.*   A power-conducting wiring path that leads from
the service panel to a specific area of the house and back to the service
panel. Some pieces of equipment used in residences draw so much
electricity that a special individual branch circuit is created just for
them. Electric ranges, dryers, refrigerators, air conditioners, and heaters
are some of the equipment that require an individual branch circuit.

*Cable.*   Electrical wiring that is covered with a protective cover or
insulating material. Romex cable, copper wires covered with heavy
paper and material for use in dry locations and covered with plastic for
use in damp locations, is the type most often used for houses. Armored
cable, BX, is sometimes used.

*Conduit.*   A thin-gaged metal pipe or tube used to protect cable
from damage.

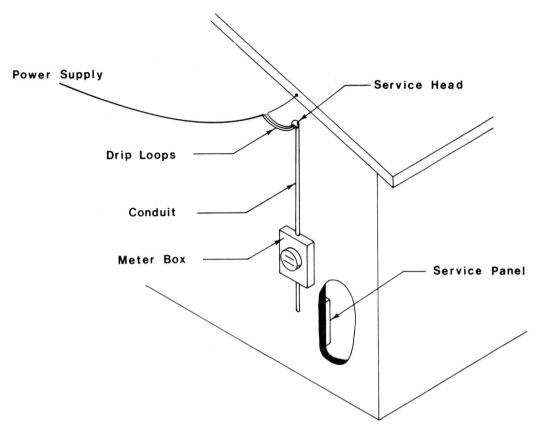

**FIGURE 11-7**   Residential wiring terms.

*Drip Loop.*   A way to suspend the service wires so as to prevent water from running down the wire into the meter or service panel.

*Mast.*   Conduit that extends above the roof of a house and is used to receive the service wires. Masts are not always used.

*Meter Box.*   A gage installed by the local power company to keep track of the amount of electricity used.

*Service Head.*   A metal cover placed on top of conduit to prevent water from entering the meter or service panel.

*Service Panel.*   A metal box that contains the main disconnect switch and circuit breakers. It also contains all connections between the main incoming power and the branch circuits.

*Service Wires.*   The power lines that bring electricity from the power company's lines to the house. The power supplied is usually 240/120-volt single-phase 60-hertz AC.

*15-amp Line, 20-amp Line, 240-volt Line.*   These terms refer to different types of branch circuits. A 15-amp line is an average line that would service a bedroom, bathroom, or similar room. A 20-amp line would be used for circuits that will probably get heavy use, such as in a kitchen where many different appliances are used. A 240-volt line is a special, very heavy duty branch circuit which is designed to carry large amounts of current. An electric hot water heater, electric range, or electric dryer would all require 240-volt lines.

## PROBLEMS

11-1   Figure P11-1 shows a floor plan of a bedroom, including a closet. Redraw the floor plan using a scale of ¼ inch = 1 foot and add, using the appropriate electrical symbols: three duplex outlets, one of which is wired to a switch; a pull switch in the closet; and a telephone jack.

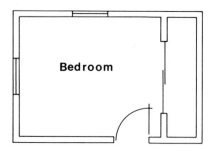

**FIGURE P11-1**   Floor plan of a bedroom. Scale: ⅛-inch = 1 foot.

11-2   Figure P11-2 shows the floor plan for a kitchen. Redraw the floor plan using a scale of ¼ inch = 1 foot and add, using the appropriate electrical symbols, each of the following:
A range outlet
A dishwasher outlet
A refrigerator outlet
An overhead light with activating switches at each door
A light over the sink, with an activating switch
A light and a fan over the range, with an activating switch for each
Four duplex outlets located over the counters
A wall outlet for a clock

**FIGURE P11-2**   Floor plan of a kitchen. Scale: ⅛-inch = 1 foot.

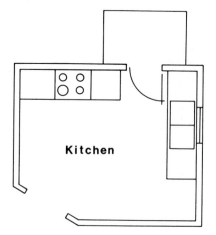

**11-3**   Figure P11-3 shows the floor plans for a two-story house. Redraw the floor plans using a scale of ¼ inch = 1 foot, and use the plans as a basis to create a set of residential electrical drawings. Think carefully about the locations of all switches, outlets, and so on, before you add them to the drawings. Redraw the floor plans, as assigned by your instructor, as follows:
   a.  First floor
   b.  Second floor
   c.  Foundation

**11-4**   Figure P11-4 shows the floor plan for a loft-type apartment. It is drawn at a scale of ⅛ inch = 1 foot. The owner has asked you to prepare an electrical plan based on the information given. Use 8½ × 11 inch paper and a scale of ¼ inch = 1 foot.

**11-5**   Figure P11-5 shows the floor plan for a house that is built into the side of a hill. It looks out over the ocean. Prepare an electrical plan. The drawing is done at ⅛ inch = 1 foot. Redraw the floor plan at a scale of ¼ inch = 1 foot. Use 11 × 17 inch or larger paper.

**FIGURE P11-3**  Floor plan of a two-story house. Scale: ⅛ -inch = 1 foot for all floors.

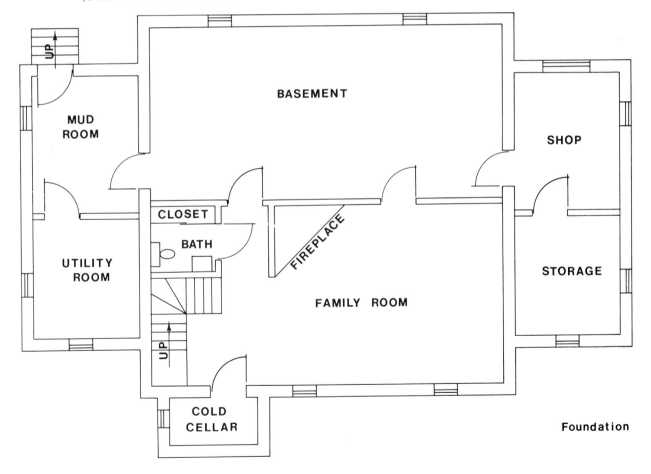

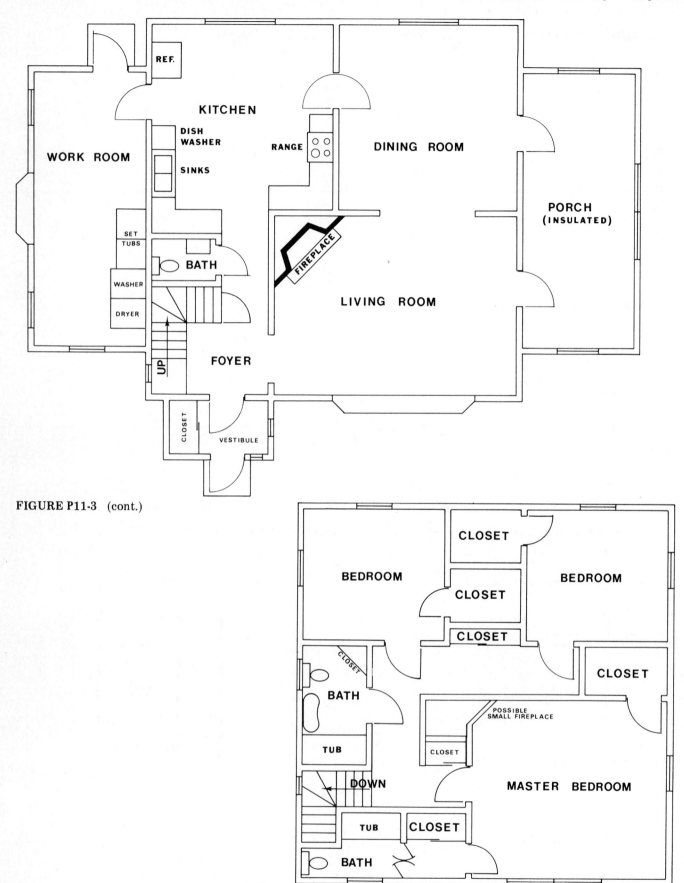

**FIGURE P11-3** (cont.)

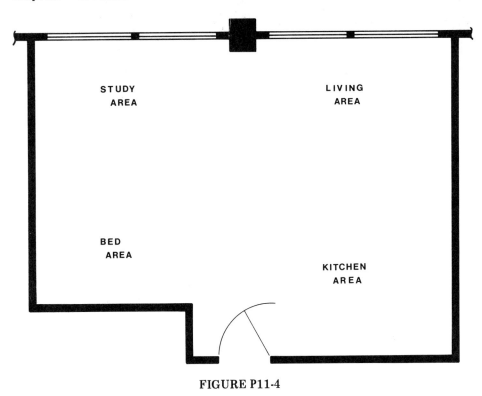

**FIGURE P11-4**

**FIGURE P11-5**

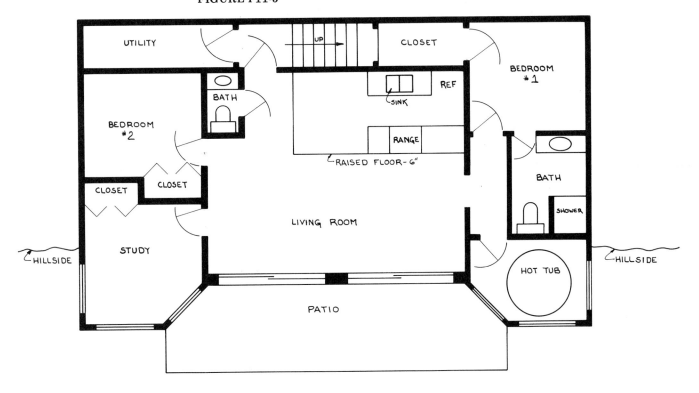

# 12

INDUSTRIAL
WIRING
DIAGRAMS

## 12-1 INTRODUCTION

In this chapter we explain how to draw some of the most widely used types of industrial wiring diagrams, including one-line diagrams, ladder diagrams, and raiser diagrams. Also presented are some of the basic design and electrical principles which these diagrams represent. The student interested in more complete coverage of the subject is referred to:

Charles W. Snow, *Electrical Drafting and Design*. Englewood Cliffs, N.J.: Prentice-Hall, Inc., 1976.

The symbols and line techniques required for each type of diagram are presented together with the text material.

## 12-2 BASIC POWER SYSTEMS

The design of any electrical system must start with the power source. What kind of power is available, and how much? For example, most houses are supplied by the local power company with power that is single-phase, three-wired, 120/240 volts, at 60 hertz. This would be expressed on a drawing as shown in Figure 12-1.

If more power is required, such as would be needed for a small apartment building, a three-phase, four-wire wye system could be used. If the power requirement is very large, as for a machine shop, a three-phase, three-wire, delta system could be used. Figure 12-2 illustrates these systems. The terms *wye* (Y) and *delta* (Δ) refer to the shape of

243

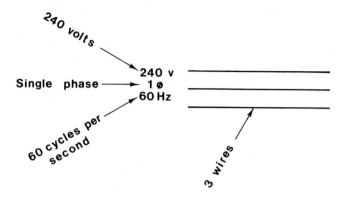

FIGURE 12-1   How to label a three-wire system.

**Three-Phase, Four-Wire Wye System**

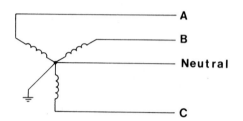

Used for apartment
buildings, office buildings,
shopping centers, etc.

**Three-Phase, Three-Wire Delta System**

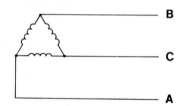

Used primarily for
factories or manufact-
uring shops.

FIGURE 12-2   Three-phase, four-wire wye system and a
three-phase, three-wire delta system.

windings used to produce the power. The basic patterns and symbols
are shown in Figure 12-3.

Figure 12-4 shows how a single-phase, three-wire system can be
used to produce different voltages. For a house, these connections are
made in the service panel. Also shown are some possible combinations
for wye and delta systems. Note that in each example, a neutral wire is
always included.

**TRANSFORMER WINDINGS**

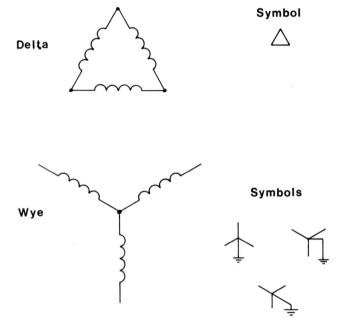

FIGURE 12-3 Transformer windings and their drawing symbols.

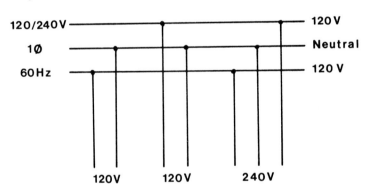

FIGURE 12-4 How a single-phase, three-wire system can be used to produce different voltages.

## 12-3 ONE-LINE DIAGRAMS

*One-line diagrams* are diagrams that graphically define the components and the relationship between the components for an electrical circuit. The components are arranged along a single vertical line which is read from top to bottom. Actually, it is rare for a one-line diagram to consist of only a single line; most often, several one-line diagrams are drawn together to show the overall power distribution system, rather than just a single circuit. Figure 12-5 shows a drawing that consists of several one-line diagrams.

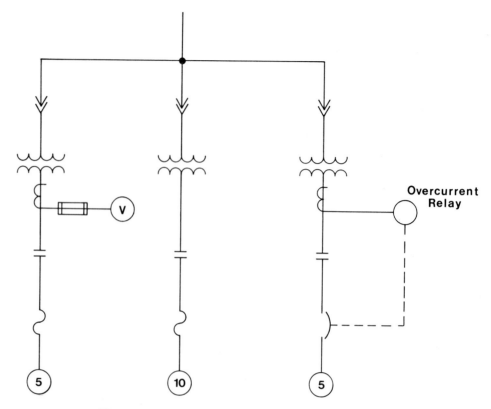

**FIGURE 12-5**  Example of three one-line diagrams which are part of the same circuit.

## 12-4  HOW TO READ
## ONE-LINE DIAGRAMS

The basic symbols used to draw one-line diagrams are presented in Figure 12-6. Figure 12-7 shows a one-line diagram that has all symbols and notations labeled. We see in Figure 12-7 that the incoming power is 480 volts, three-phase, 60 Hz. The power passes through the connection and down the line that represents the circuit. The power then passes through a transformer which has a delta winding on one side and a grounded wye winding on the other. The resulting current is 120 volts, one phase, 60 Hz. Next, an ammeter is placed on the line. The power rating and range of the ammeter are defined. An overload protection device is located in the ammeter line to protect the ammeter from damage.

The main circuit then continues through a circuit breaker, starter switch, another overload protection device, and finally into the 5-horsepower electrical motor at the end of the line.

It should be noted that the labeling on a one-line diagram is an important aspect of the diagram. The drafter should always carefully check *all* labeling used and then have the design engineer double-check the labeling.

# SYMBOLS for ONE LINE DIAGRAMS

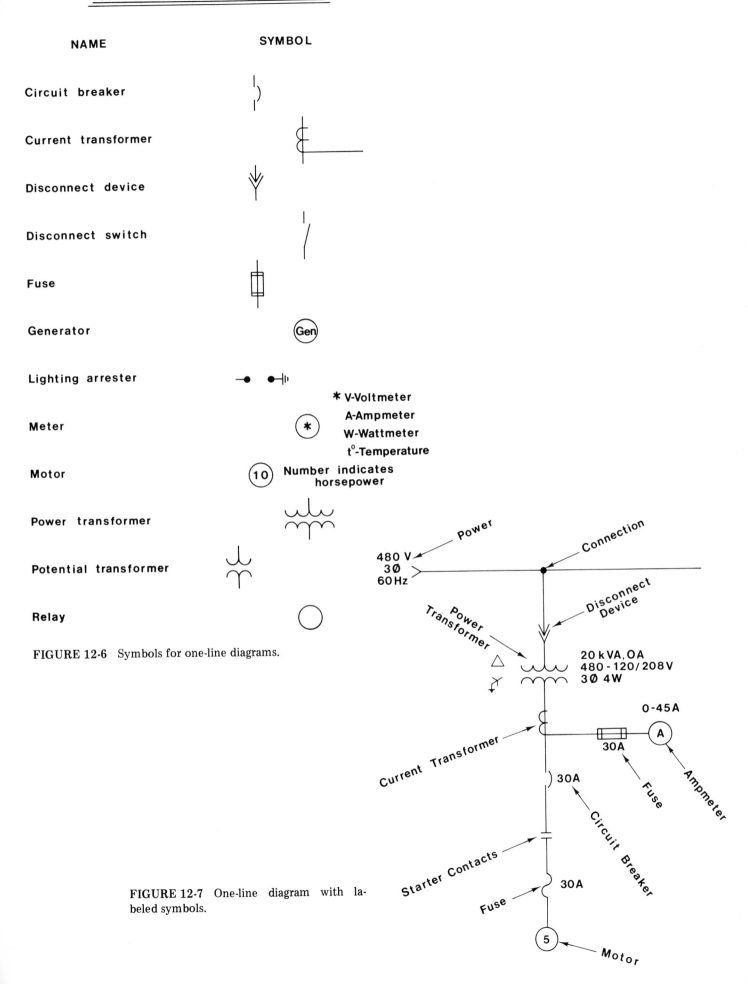

| NAME | SYMBOL |
|------|--------|
| Circuit breaker | |
| Current transformer | |
| Disconnect device | |
| Disconnect switch | |
| Fuse | |
| Generator | Gen |
| Lighting arrester | |
| Meter | * V-Voltmeter  A-Ampmeter  W-Wattmeter  t°-Temperature |
| Motor | 10 Number indicates horsepower |
| Power transformer | |
| Potential transformer | |
| Relay | |

FIGURE 12-6  Symbols for one-line diagrams.

Power

480 V
3Ø
60 Hz

Connection

Power Transformer

Disconnect Device

20 kVA, OA
480 - 120/208 V
3Ø 4W

0-45A

A

30A

Current Transformer

30A

Fuse

Ampmeter

30A

Circuit Breaker

Starter Contacts

Fuse    30A

5

Motor

FIGURE 12-7  One-line diagram with labeled symbols.

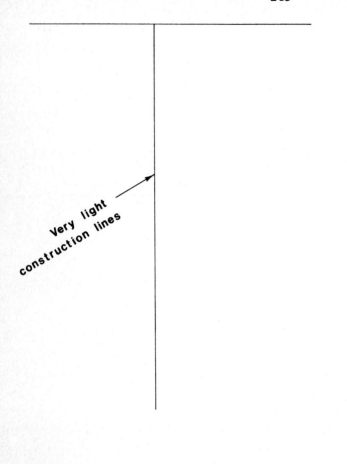

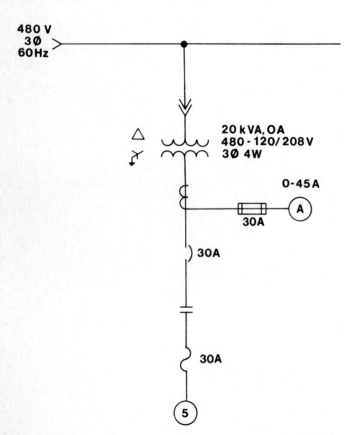

480 V
3Ø
60 Hz

20 kVA, OA
480 - 120/208 V
3Ø 4W

0-45A

30A

A

30A

30A

5

Very light
construction lines

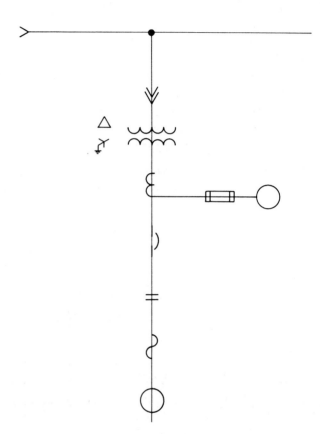

**FIGURE 12-8** How to draw a one-line diagram.

## 12-5 HOW TO DRAW
## ONE-LINE DIAGRAMS

To draw a one-line diagram, use the following procedure (illustrated in Figure 12-8):

1. Prepare a freehand sketch of the entire circuit and check all symbols and labeling.
2. Draw very light horizontal lines to represent the power source and a vertical line as shown.
3. Add all symbols for the various components.
4. Add guidelines for labeling.
5. Darken all lines to their final color and configuration. Add labeling.

## 12-6 LADDER DIAGRAMS

Ladder diagrams are a type of diagram used to define industrial control circuits. They derive their name from the fact that they resemble ladders—two long parallel vertical lines connected by a series of horizontal runners. Figure 12-9 is an example of a ladder diagram.

Ladder diagrams are most commonly used to draw the circuitry required to activate and deactivate electrical motors and coils. These components, in turn, are used to operate elevators, heating and air-conditioning systems, machines, subway systems, and so on.

**FIGURE 12-9** Example of a ladder diagram.

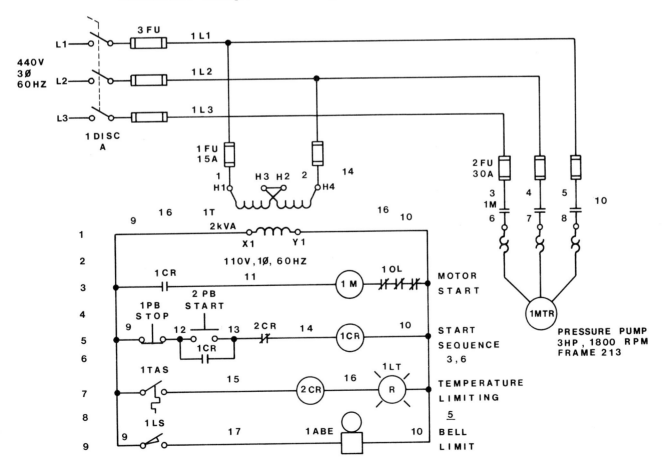

## 12-7  HOW TO READ LADDER DIAGRAMS

Ladder diagrams are drawn in two parts: the horizontal upper portion, which defines the power applications; and the vertical/horizontal ladder portion, which defines the control functions. The diagram is read from left to right across the power lines, then down the left side of the ladder portion and across each individual horizontal line from left to right. Figure 12-10 shows this concept.

The symbols and abbreviations used on ladder diagrams are presented in Figure 12-11. Not only are the symbols important when drawing a ladder diagram, but the labeling of the symbols is equally important. Ladder diagrams usually include many switches and coils which, from a symbol viewpoint, appear exactly the same. The only way to tell which switch is related to which coil is by the labeling. For example, in Figure 12-9 CR1 activates switch CR1 and the coil labeled CR2 activates CR2.

Switches are always drawn in their "natural" positions, that is, in the position they are in when not activated by current. A normally open switch is one that is open unless closed by current and is therefore drawn in the open position.

Figure 12-12 has been prepared to demonstrate how to read ladder diagrams. It is a ladder diagram showing the circuitry needed to start and stop an electric motor. The current enters the circuit at the positive terminal and exists at the negative terminal. Each part of the diagram has been labeled so that you can follow the current as it flows through the circuit. Note that the diagram, as drawn, shows all switches in their natural positions and that there are no completed paths across the diagram. Only when the start switch is pressed is there a completed path

**FIGURE 12-10**  Ladder diagrams are read from left to right and top to bottom.

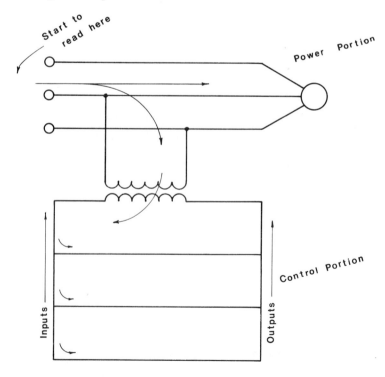

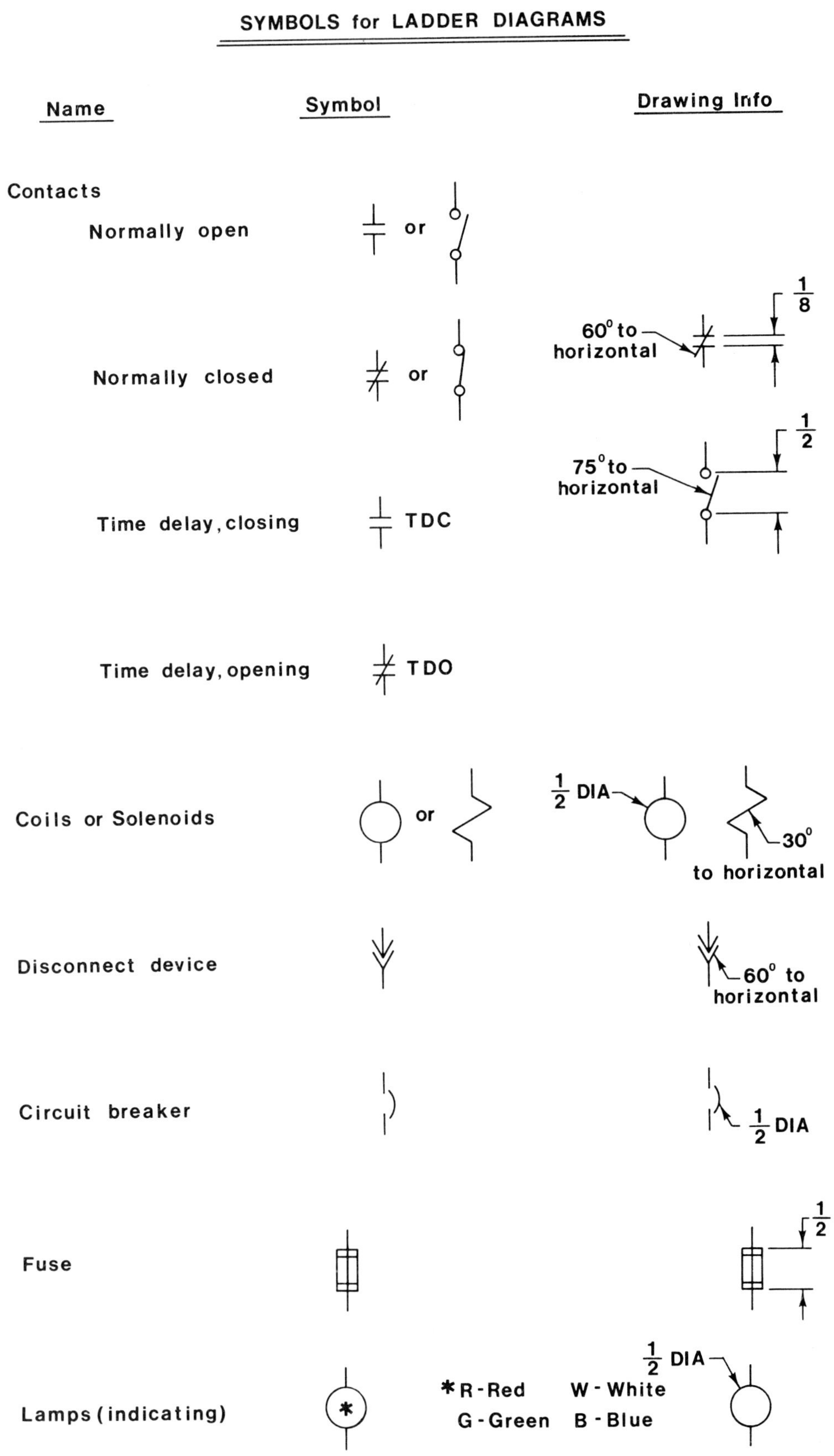

## SYMBOLS for LADDER DIAGRAMS

| Name | Symbol | Drawing Info |
|---|---|---|

**Contacts**

    Normally open

    Normally closed

    Time delay, closing    TDC

    Time delay, opening    TDO

**Coils or Solenoids**

**Disconnect device**

**Circuit breaker**

**Fuse**

**Lamps (indicating)**    *R - Red    W - White    G - Green    B - Blue

FIGURE 12-11  Symbols for ladder diagrams.

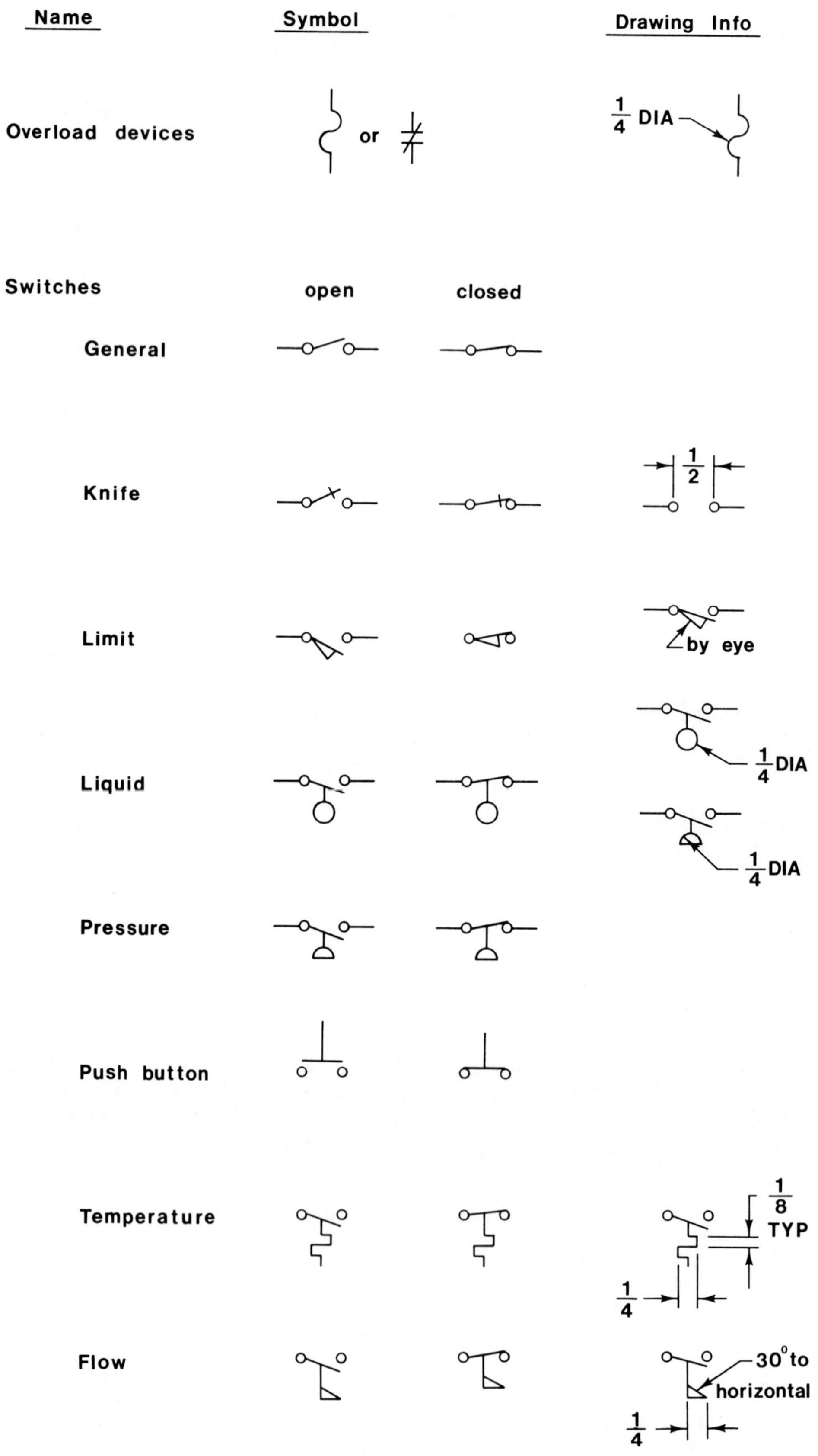

Name
Symbol
Drawing Info
Overload devices
or
1/4 DIA
Switches
open
closed
General
Knife
1/2
Limit
by eye
Liquid
1/4 DIA
1/4 DIA
Pressure
Push button
Temperature
1/8 TYP
1/4
Flow
30° to horizontal
1/4

FIGURE 12-11 (cont.)

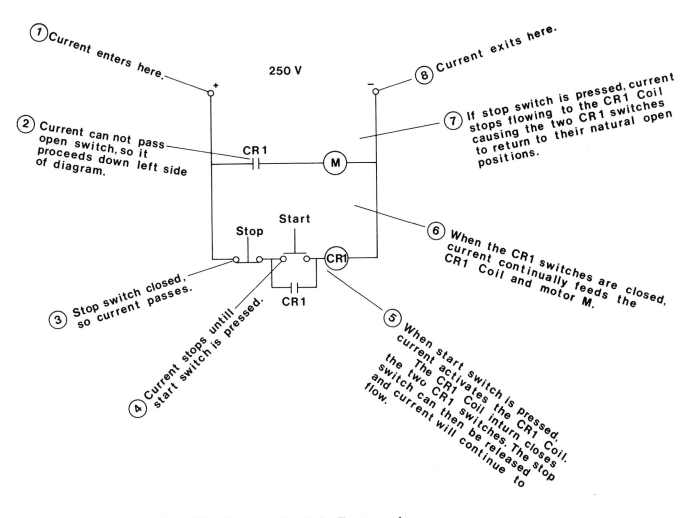

**① Current enters here.**

**250 V**

**② Current can not pass open switch, so it proceeds down left side of diagram.**

**⑧ Current exits here.**

**⑦ If stop switch is pressed, current stops flowing to the CR1 Coil causing the two CR1 switches to return to their natural open positions.**

**③ Stop switch closed, so current passes.**

**④ Current stops untill start switch is pressed.**

**⑥ When the CR1 switches are closed, current continually feeds the CR1 Coil and motor M.**

**⑤ When start switch is pressed, current activates the CR1 Coil. The CR1 Coil inturn closes the two CR1 switches. The stop switch can then be released and current will continue to flow.**

CR1

M

Stop    Start

CR1

CR1

FIGURE 12-12   Ladder diagram with labels. Try to read the story the symbols tell.

which permits current to flow. Once current is flowing, and the CR1 coil is activated, the CR1 switches close and the start switch may be released.

All circuits should include overload protection. This may be in the form of fuses or temperature- or current-sensitive switches. These devices are also drawn in their natural position, which in most cases is the closed position. The standard abbreviation for overload protection is OL.

## 12-8   HOW TO DRAW A LADDER DIAGRAM

The procedure used to create ladder diagrams is as follows (see Figure 12-13):

1.  Make a freehand sketch of the entire circuit. A freehand sketch is helpful for two reasons: (1) it enables you to work out in advance the entire circuit, thereby reducing your chances of making an error in the final drawing; and (2) it enables you to judge the

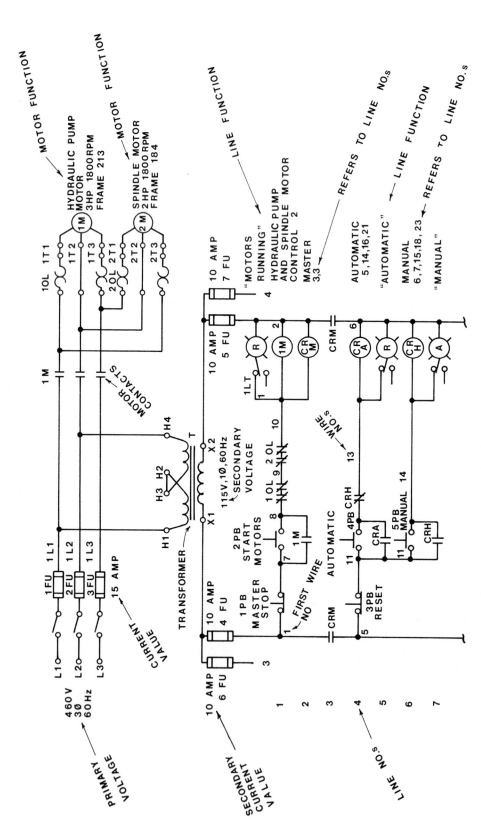

FIGURE 12-13  How to draw a ladder diagram.

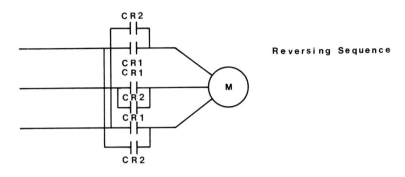

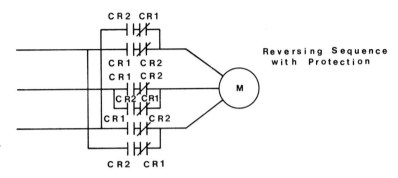

**FIGURE 12-14**   Reversing sequence.

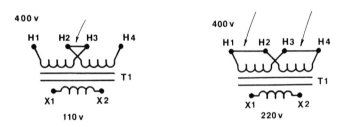

**FIGURE 12-15**   A transformer 440 volts to 220 volts.

approximate space requirements of the diagram. Remember that it is much easier to change a sketch than to change a final drawing.

2. Draw in the horizontal power lines. Use light construction lines, as parts of the lines will be erased. If two motors are required, use the arrangement shown in Figure 12-13. If a reversing sequence is required, use the setup shown in Figure 12-14.

3. Add the disconnect switches, fuses, and starter contacts as shown. The overload devices shown near the motor are usually manufactured as part of the motor. Nevertheless, they must be included on the diagram.

4. Add the transformer. Transformers that convert 440 volts to 110 volts have terminals H3 and H2 connected as shown in Figure 12-13. Transformers that convert 440 volts to 220 volts connect pins H1 to H3 and H2 to H4, as shown in Figure 12-15.

5. Lay out the control portion of the diagram. There is no standard spacing, but placing the vertical lines 6 inches apart and the horizontal lines 1 inch apart will usually be sufficient. Larger diagrams require larger spacing.

6. Add the required symbols and label the diagram.

**12-9    HOW TO LABEL
A LADDER DIAGRAM**

Figure 12-13 shows a labeled ladder diagram. Each label has been identified by a note. Specific labeling will vary from company to company depending on the product being made and individual preferences.

The diagram used is part of an example presented by the Joint International Council (JIC) of 7901 Westpark Drive, McLean, Va. 22102 in their publication *Electrical Standards for Mass Production Equipment* EMP-1-67. The JIC standards are generally accepted and used throughout the electrical industry.

**12-10    RISER DIAGRAMS**

Riser diagrams are drawings that show the wiring paths of a building's electrical system up to, but not including, the branch circuits. Figure 12-16 is an example of a riser diagram which shows the between-floor wiring paths of a security system. Note that all lines drawn are uniform in intensity and thickness and that all lettering is as outlined in Figure 1-9. The size of the blocks may be varied according to individual needs, but the distance between floor lines is generally kept equal.

**12-11    HOW TO READ A RISER DIAGRAM**

Riser diagrams are read by starting at the power input point and then following each wiring path to its destination. In Figure 12-16 we see that the power enters the building in the basement. It then proceeds to the first security center console and from there to the individual stations.

The legend in the lower right-hand corner of the drawing defines all the symbols used. All symbols should be clearly defined even if their meaning seems to be obvious.

Each floor must be labeled and separated by a horizontal line. Special or large important components such as the security center console and the main power console are labeled on the drawing as shown. They should not be abbreviated.

**12-12    HOW TO DRAW A RISER DIAGRAM**

Figure 12-17 illustrates how to create a riser diagram. The diagram was originally presented in sketch form in Figure 12-18. The procedure used is as follows:

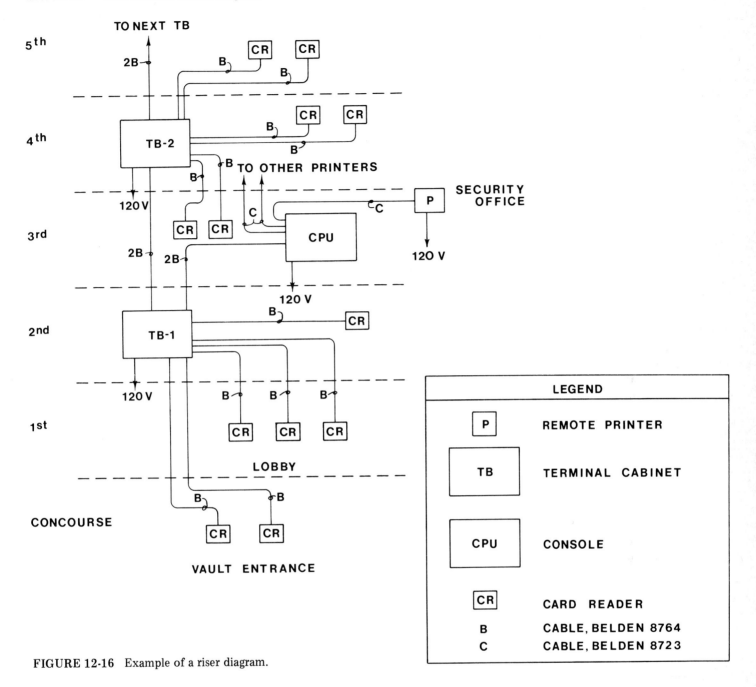

FIGURE 12-16    Example of a riser diagram.

1.  Study the information and requirements to determine how many floors are to be drawn and what components are needed. Draw the basic floor pattern so that floors are evenly spaced. If a floor has a large number of components, its size may be increased.

2.  Draw in all major components.

3.  Draw in all minor components and add wiring paths.

4.  Label all components and set up a legend. Every component used must be defined in the legend.

5.  Darken all lines to their final color and configuration.

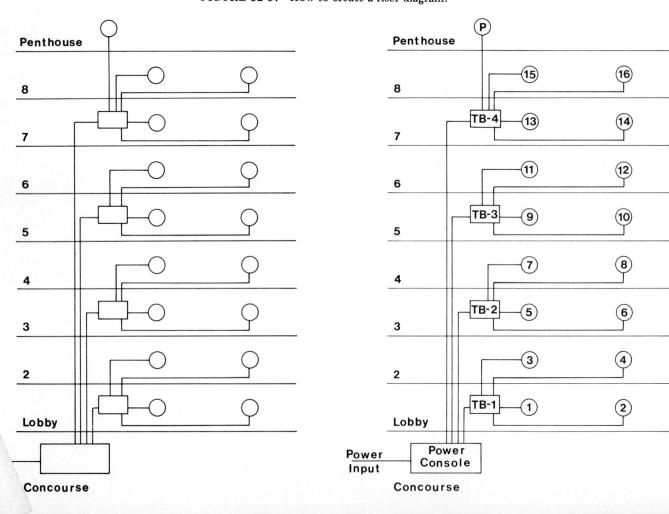

**FIGURE 12-17** How to create a riser diagram.

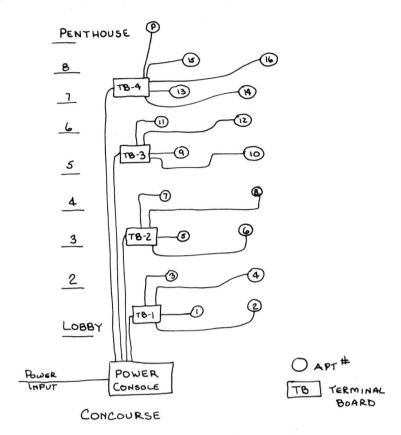

**FIGURE 12-18**  Freehand sketch of a riser diagram from which Figure 12-15 was prepared.


## PROBLEMS

12-1  Prepare a one-line diagram of the sketch shown in Figure P12-1.

12-2  Prepare a one-line diagram of the sketch shown in Figure P12-2.

12-3  Prepare a one-line diagram of the sketch shown in Figure P12-3.

12-4  Redraw the ladder diagram shown in Figure P12-4 and substitute the following symbols:
   1. Fuse L1
   2. Fuse L2
   3. Contact, normally open, CR1
   4. A motor M
   5. An overload device, OL
   6. Transformer, 115 V to 60 V
   7. An overload device, OL
   8. A start/stop sequence which includes a start button, a stop button, a coil labeled CR1, and a normally open switch labeled CR1
   9. An overload device, OL

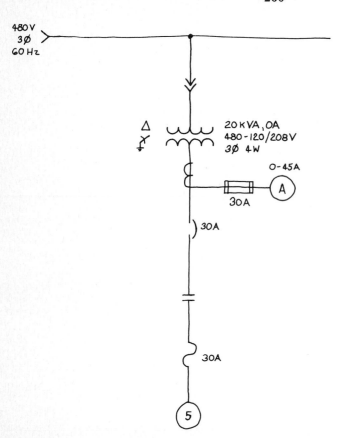

480V
3∅
60Hz

△
Y

20 KVA , OA
480 - 120/208 V
3∅ 4W

O- 45A

A

30A

30A

30A

30A

5

FIGURE P12-1

5 KV

225A

225A

5KV-220V

CONTROL
CIRCUIT

5KV-220V

CONTROL
CIRCUIT

2-600/5

OL

2-200/5

OL

M

M

A

REAC

MOT
NO I

2500 HP

MOT
NO 2

1000 HP

FIGURE P12-2

FIGURE P12-3

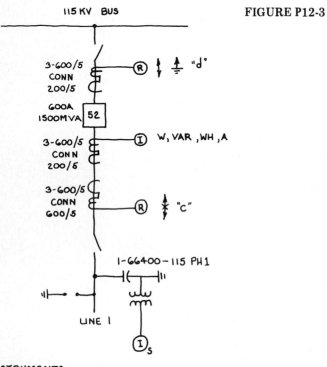

115 KV   BUS

3-600/5
CONN
200/5

R  ↕ ⏚  "d"

600A
1500MVA  52

3-600/5
CONN
200/5

I   W, VAR ,WH ,A

3-600/5
CONN
600/5

R  ↕  "c"

1-66400-115 PH 1

LINE 1

I S

FIGURE P12-4

1

2

3      4      5

6

7                    9

8

I   INSTRUMENTS

R   RELAY

"d"  DIR GRD

"c"  BUS DIFF

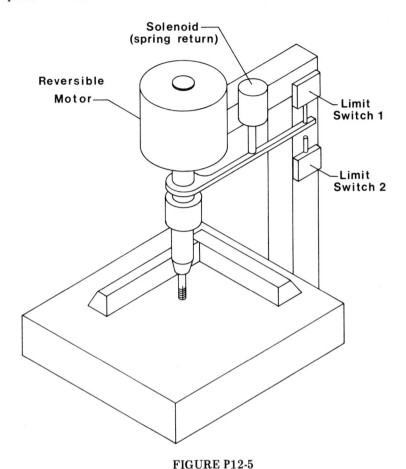

**FIGURE P12-5**

12-5   Figure P12-5 shows a drill press that has been rigged for automatic operation. The operator positions the workpiece under
the drill and pushes a start button. The drill then follows the
sequence listed.
1.  Motor starts—forward direction.
2.  Solenoid activates and pushes drill downward.
3.  Limit switch 2 is pressed.
4.  Solenoid deactivates and returns (spring-loaded return).
5.  Motor stops and then reverses.
6.  Limit switch 1 is pressed.
7.  The motor stops and the entire system is stopped.
8.  All current is shut off.
Prepare a ladder diagram that represents the drilling sequence.
Be sure to include overload and safety protection devices.

12-6   This problem is exactly like Problem 12-5 except that in this
case two more solenoids are added as clamping devices. These
solenoids will press the workpiece against the guide rails shown
and hold the workpiece during drilling. After the drilling is
complete, the solenoids will release the workpiece.

**12-7**  Redraw the riser diagram shown in Figure P12-7 and add a fourth floor with two more alarms.

### WHEATLEY TRUST CO.
### ALARM SYSTEM

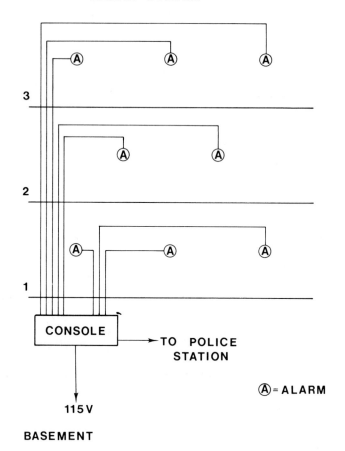

FIGURE P12-7

**12-8**  Redraw the riser diagram shown in Figure P12-8.

**12-9**  Redraw the riser diagram shown in Figure P12-9.

**12-10**  Redraw Figure 12-8 approximately twice as large as presented.

**12-11**  Redraw Figure 12-12 approximately twice as large as presented.

**12-12**  Design a circuit that will raise and lower the dumbwaiter as shown in Figure P12-12. The dumbwaiter must be able to be called from either station.

**12-13**  Repeat Problem 12-12 but include automatic door openers and a safety system that prevents a person's hand from being jammed in the door.

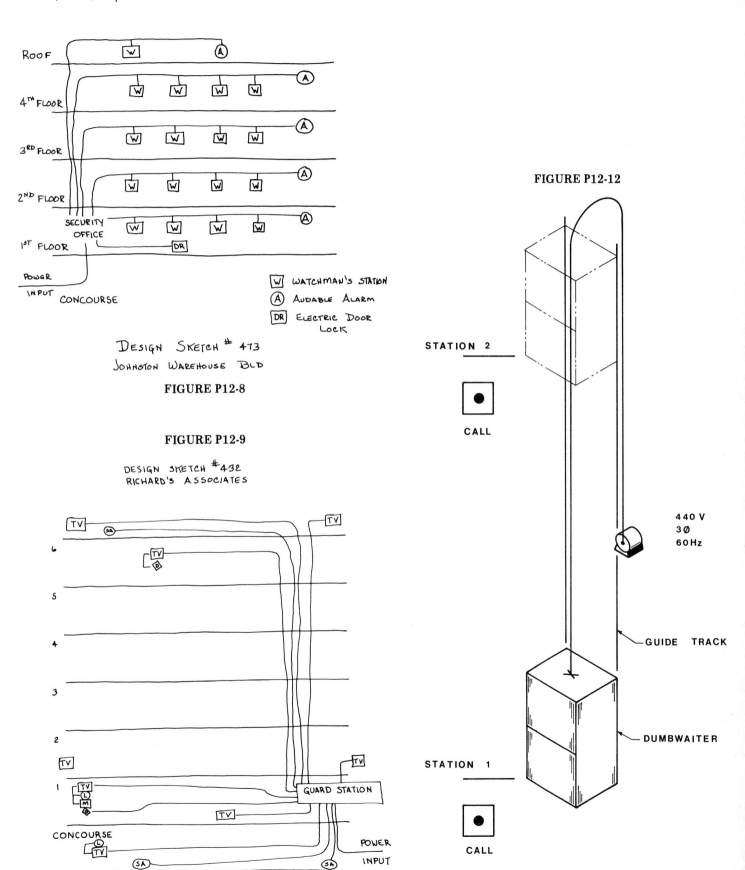

RISER DIAGRAM — WATCHMAN'S TOUR

ROOF

4TH FLOOR

3RD FLOOR

2ND FLOOR

SECURITY OFFICE

1ST FLOOR

POWER INPUT

CONCOURSE

W  WATCHMAN'S STATION
A  AUDABLE ALARM
DR  ELECTRIC DOOR LOCK

DESIGN SKETCH # 473
JOHNSTON WAREHOUSE BLD

**FIGURE P12-8**

**FIGURE P12-9**

DESIGN SKETCH # 432
RICHARD'S ASSOCIATES

CONCOURSE

POWER INPUT

GUARD STATION

SUB CONCOURSE

LEDGERD          SA  SILENT ALARM
TV  T.V. CAMERA    L  LISTENING DEVICE    M  LOUDSPEAKER    D  DOOR LOCK

**FIGURE P12-12**

STATION 2

CALL

440 V
3 Ø
60 Hz

GUIDE TRACK

DUMBWAITER

STATION 1

CALL

# 13

COMPUTER
APPLICATIONS
TO ELECTRONIC
AND ELECTRICAL
DRAFTING

## 13-1   INTRODUCTION

Computers have made a tremendous impact on electrical and electronic drafting. Computer systems enable drafters to create drawings more quickly than can be done using conventional pencil techniques. Lines and symbols can be completely erased or moved around the drawing using simple commands.

Figure 13-1 shows a typical computer graphics setup. Systems vary but, in general, use the same operating format. All systems use a viewing screen called a visual display unit (VDU). In addition, all systems have a keyboard which is used to input commands to the computer. Some systems have a keyboard and a menu board which can also be used to send commands to the computer. A keyboard is used to send coded commands, usually words, whereas a menu board contains a visual listing of symbols and sends a complete command in one input. For example (see Figure 13-2), if we wanted to draw an AND symbol, we might type in the word "AND" or some code word, or we could simply touch the appropriate section on the menu board.

It should be noted that most systems do not require the user to be a computer programmer. The user simply learns the appropriate commands for the system being used.

## 13-2   HOW TO CREATE A DRAWING

The first step in creating a drawing using a computer system is to call for a grid pattern to appear on the display screen. Depending on the system, the grid will appear as dots, crosses, or broken lines (see Figure 13-3).

**TYPICAL COMPUTER
GRAPHICS SETUP**

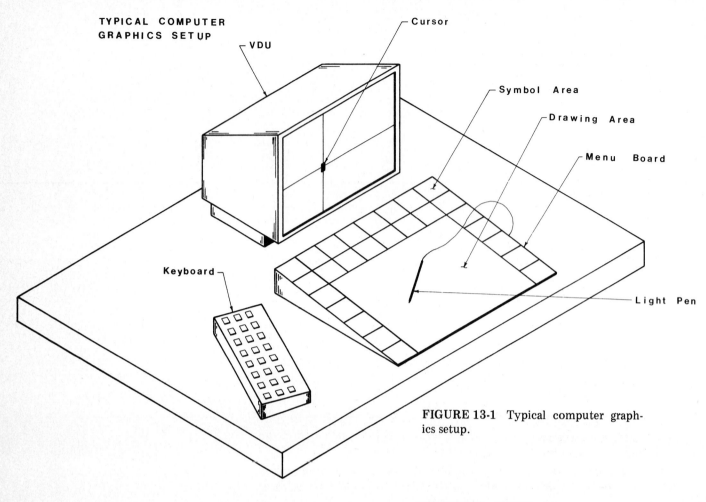

FIGURE 13-1   Typical computer graphics setup.

**DIFFERENT TYPES OF
INPUT COMMANDS**

**FIGURE 13-2** How to draw an AND symbol.

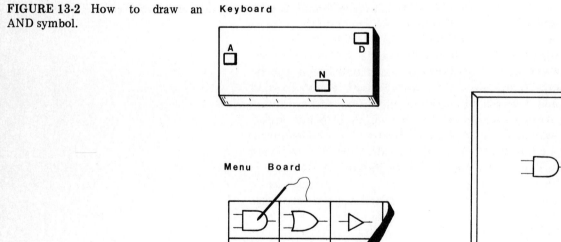

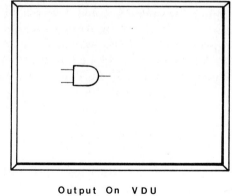

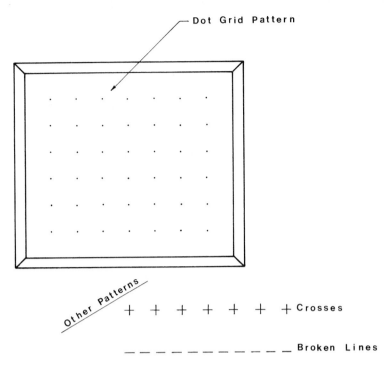

FIGURE 13-3   How a grid pattern is displayed on a computer screen.

The grid spacing may be set up in inches or in metric units. The grid is used just as it is used for preparing conventional pencil drawings: to space and align the symbols and conductor paths.

Symbols are called for by using either word codes via the keyboard or by using a light pen and a menu board, as shown in Figure 13-2.

When the light pen is touched on the appropriate symbol of the menu board, the symbol will appear on the screen. The symbol will usually align with the cursor. The cursor is a small square dot that appears on the screen and is used to locate and move information. The cursor is moved, thereby moving the symbol, by moving the light pen around with the drawing reference area (see Figure 13-4). Other systems use thumb screws or a small rectangular object called a "mouse." Once the symbol is located, another command is used to add the symbol to the drawing. Lines are added in a similar manner.

Symbols can be rotated using angular inputs. For example, Figure 13-5 shows an AND gate in four possible positions: $0°$, $90°$, $180°$, and $270°$. The counterclockwise direction is always considered positive, as shown in Figure 13-6. This means that $30°$ clockwise must be stated as $300°$ counterclockwise.

## 13-3   TYPES OF SYSTEMS AVAILABLE

There are many different types of systems available. Some, like the Data General system shown in Figure 13-7, include their own computer and can be programmed to prepare many different types of drawings. A system like the Apollo, shown in Figure 13-8, uses an Apollo mainframe computer but software packages designed by other companies.

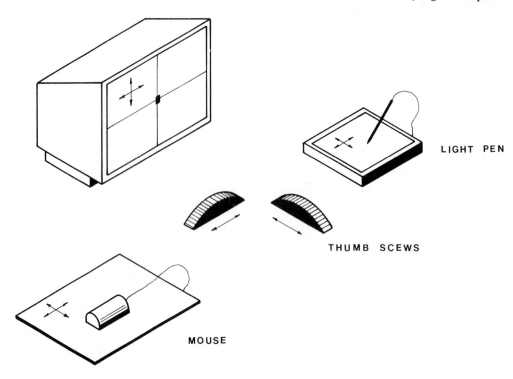

**FIGURE 13-4**    Various types of systems used to control a cursor.

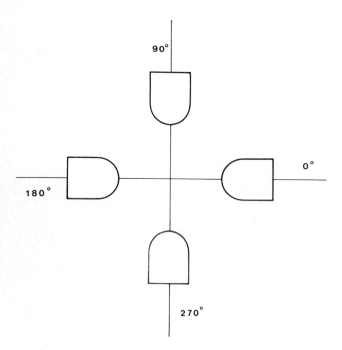

**FIGURE 13-5**    Various positions for an AND gate.

**FIGURE 13-6**    Computer rotates only in the counterclockwise direction.

ALL ANGLES ARE MEASURED
IN THIS DIRECTION

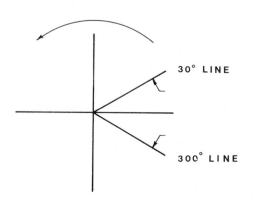

FIGURE 13-7 Example of a computer graphic system. *(Courtesy of Data General, Westboro, Mass.)*

FIGURE 13-8 Can be used with an Apollo system. (*Courtesy of Mentor Corp.*)

The computer company together with the associated software companies constitute a *domain system*.

In computer terminology, *hardware* refers to the actual computer. *Software* refers to the programs that operate the computer. *Canned programs* are prepared programs that require simple input commands from the user.

## PROBLEMS

Any of the problems presented in previous chapters may be done using a computer graphic system. The following are a suggested general list.

13-1    Problem 2-6
13-2    Problem 3-4
13-3    Problem 3-14
13-4    Problem 4-5
13-5    Problem 5-20
13-6    Problem 5-22
13-7    Problem 6-1
13-8    Problem 6-5
13-9    Problem 10-4
13-10   Problem 11-3
13-11   Problem 11-7

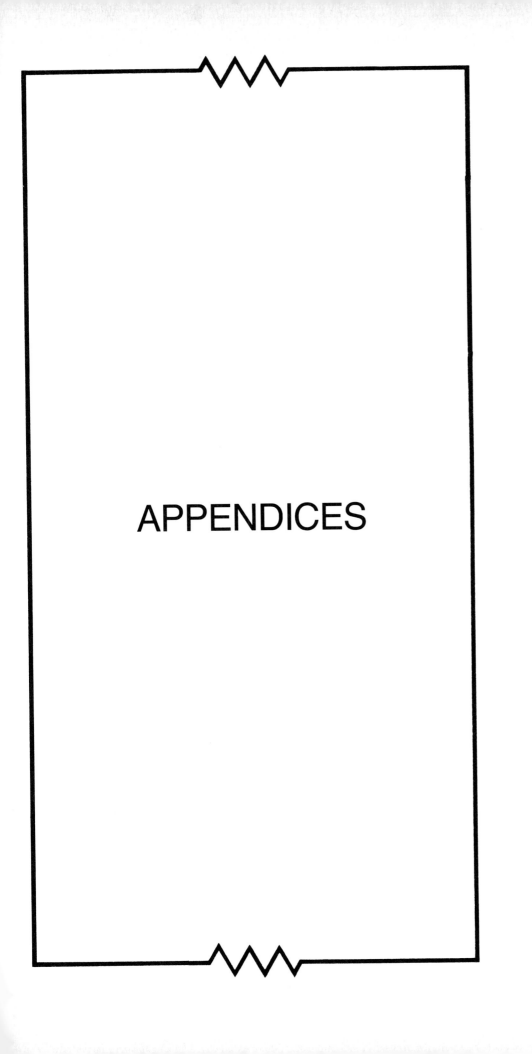

APPENDICES

# A DIGITAL READOUT LETTERS

**DIGITAL READOUTS**

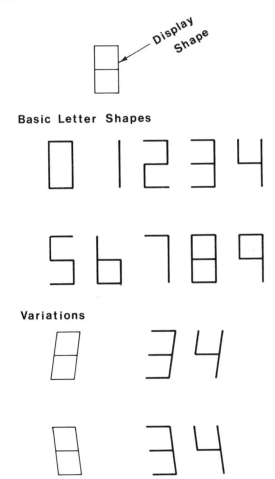

Basic Letter Shapes

Variations

# B RESISTOR COLOR CODES

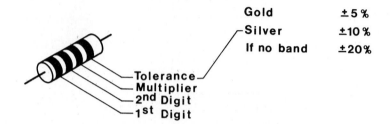

| | | |
|---|---|---|
| Gold | ±5% |
| Silver | ±10% |
| If no band | ±20% |

Tolerance
Multiplier
2nd Digit
1st Digit

| Color | 1st Digit | 2nd Digit | Multiplier |
|---|---|---|---|
| Black | 0 | 0 | 1 |
| Brown | 1 | 1 | 10 |
| Red | 2 | 2 | 100 |
| Orange | 3 | 3 | 1,000 |
| Yellow | 4 | 4 | 10,000 |
| Green | 5 | 5 | 100,000 |
| Blue | 6 | 6 | 1,000,000 |
| Violet | 7 | 7 | 10,000,000 |
| Gray | 8 | 8 | 100,000,000 |
| White | 9 | 9 | 1,000,000,000 |
| Gold | – | – | .1 |
| Silver | – | – | .01 |

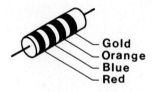

Gold
Orange
Blue
Red

2600 ohms ±5%

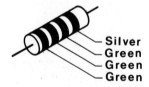

Silver
Green
Green
Green

550,000 ohms ±10%

# C WIRE AND SHEET METAL GAGES

| WIRE AND SHEET METAL GAGES | | | |
|---|---|---|---|
| Gage | Thickness | Gage | Thickness |
| 000 000 | 0.5800 | 18 | 0.0403 |
| 00 000 | .5165 | 19 | .0359 |
| 0 000 | .4600 | 20 | .0320 |
| 000 | .4096 | 21 | .0285 |
| 00 | .3648 | 22 | .0253 |
| 0 | .3249 | 23 | .0226 |
| 1 | .2893 | 24 | .0201 |
| 2 | .2576 | 25 | .0179 |
| 3 | .2294 | 26 | .0159 |
| 4 | .2043 | 27 | .0142 |
| 5 | .1819 | 28 | .0126 |
| 6 | .1620 | 29 | .0113 |
| 7 | .1443 | 30 | .0100 |
| 8 | .1285 | 31 | .0089 |
| 9 | .1144 | 32 | .0080 |
| 10 | .1019 | 33 | .0071 |
| 11 | .0907 | 34 | .0063 |
| 12 | .0808 | 35 | .0056 |
| 13 | .0720 | 36 | .0050 |
| 14 | .0641 | 37 | .0045 |
| 15 | .0571 | 38 | .0040 |
| 16 | .0508 | 39 | .0035 |
| 17 | .0453 | 40 | .0031 |

# D PAPER SIZES

## PAPER SIZES

| Size | Dimensions |
|------|------------|
| A | $8\frac{1}{2}$ x 11 |
| B | 11 x 17 |
| C | 17 x 22 |
| D | 22 x 34 |
| E | 34 x 44 |
| J | Roll Size |

# E WIRE COLOR CODES

| COLOR CODES | | |
|---|---|---|
| COLOR | LETTER CODE | NUMBER CODE |
| Black | BK | 0 |
| Brown | BR | 1 |
| Red | R | 2 |
| Orange | O | 3 |
| Yellow | Y | 4 |
| Green | GR | 5 |
| Blue | BL | 6 |
| Violet | V | 7 |
| Gray | GY | 8 |
| White | W | 9 |

# F STANDARD SCREW THREADS

| NOMINAL DIA | | UNC | | UNF | | UNEF | |
|---|---|---|---|---|---|---|---|
| | | PITCH | TAP | PITCH | TAP | PITCH | TAP |
| No.1 | .073 | 64 | No. 53 | 72 | No. 53 | | |
| 2 | .086 | 56 | 50 | 64 | 50 | | |
| 3 | .099 | 48 | 47 | 56 | 45 | | |
| 4 | .112 | 40 | 43 | 48 | 42 | | |
| 5 | .125 | 40 | 38 | 44 | 37 | | |
| 6 | .138 | 32 | 36 | 40 | 33 | | |
| 8 | .164 | 32 | 29 | 36 | 29 | | |
| 10 | .190 | 24 | 25 | 32 | 21 | | |
| No.12 | .216 | 24 | 16 | 28 | 14 | 32 | No. 13 |
| $\frac{1}{4}$ | .250 | 20 | No. 7 | 28 | No. 3 | 32 | No. 2 |
| $\frac{5}{16}$ | .313 | 18 | F | 24 | I | 32 | K |
| $\frac{3}{8}$ | .375 | 16 | $\frac{5}{16}$ | 24 | Q | 32 | S |
| $\frac{7}{16}$ | .438 | 14 | U | 20 | $\frac{25}{64}$ | 28 | Y |
| $\frac{1}{2}$ | .500 | 13 | $\frac{27}{64}$ | 20 | $\frac{29}{64}$ | 28 | $\frac{15}{32}$ |

# G STANDARD METRIC THREADS

| NOMINAL DIA [mm] | COARSE | | FINE | |
|---|---|---|---|---|
| | PITCH | TAP | PITCH | TAP |
| 1.6 | 0.35 | 1.25 | | |
| 2 | 0.40 | 1.6 | | |
| 2.5 | 0.45 | 2.05 | | |
| 3 | 0.5 | 2.5 | | |
| 4 | 0.7 | 3.3 | | |
| 5 | 0.8 | 4.2 | | |
| 6 | 1 | 5.0 | | |
| 8 | 1.25 | 6.7 | 1 | 7.0 |
| 10 | 1.5 | 8.5 | 1.25 | 8.0 |
| 12 | 1.75 | 10.2 | 1.25 | 10.8 |
| 16 | 2 | 14 | 1.5 | 14.5 |
| 20 | 2.5 | 17.5 | 1.5 | 18.5 |

NOTE: OTHER SIZES ARE AVAILABLE
BUT THOSE LISTED ARE "PREFERRED"
FOR GENERAL USE

# H STANDARD COMPONENT SIZES

THESE DIMENSIONS ARE TYPICAL SIZES.
MANY OTHER SIZES ARE AVAILABLE

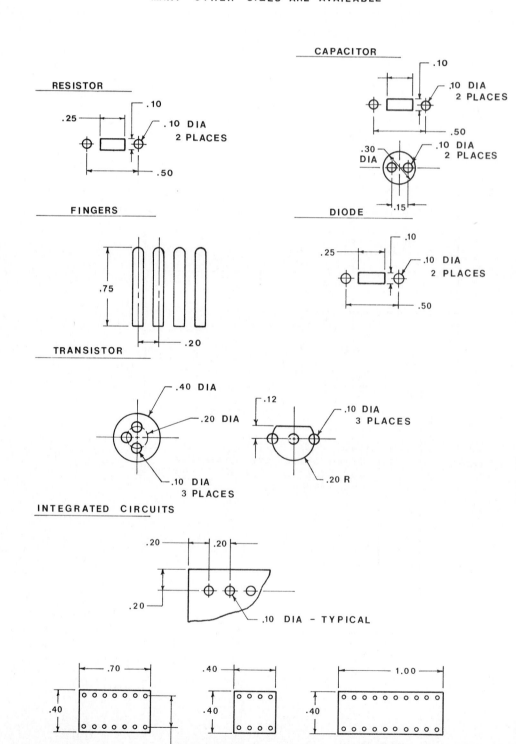

RESISTOR

CAPACITOR

FINGERS

DIODE

TRANSISTOR

INTEGRATED CIRCUITS

# LIGHT
# BOX

LIGHT BOX

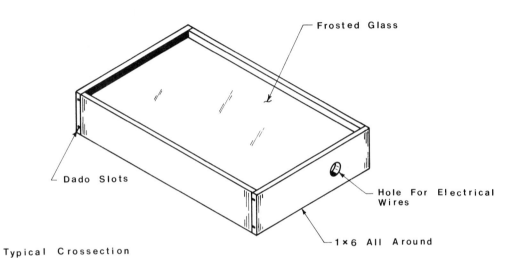

Frosted Glass

Dado Slots

Hole For Electrical Wires

1×6 All Around

Typical Crossection

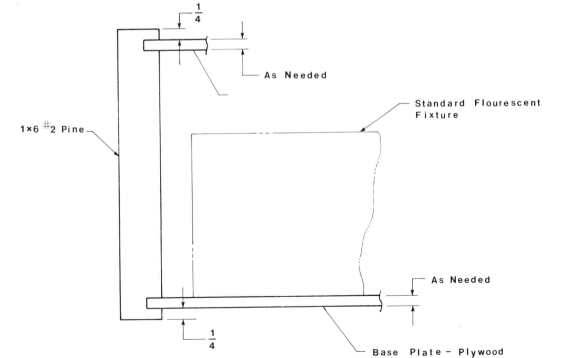

$\frac{1}{4}$

As Needed

Standard Flourescent Fixture

1×6 #2 Pine

As Needed

$\frac{1}{4}$

Base Plate - Plywood

Cut Lengths To Fit
Flourescent Fixture

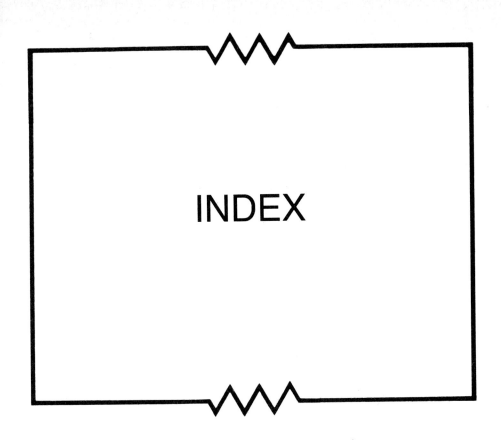

# INDEX